Secrets to Success in Industry Careers

SECRETS TO SUCCESS IN INDUSTRY CAREERS

Essential skills for science and business

L. Borbye

ELSEVIER

AMSTERDAM • BOSTON • HEIDELBERG • LONDON • NEW YORK • OXFORD
PARIS • SAN DIEGO • SAN FRANCISCO • SINGAPORE • SYDNEY • TOKYO
Academic Press is an imprint of Elsevier

Academic Press is an imprint of Elsevier
84 Theobald's Road, London WC1X 8RR, UK
30 Corporate Drive, Suite 400, Burlington, MA 01803, USA
525 B Street, Suite 1900, San Diego, CA 92101-4495, USA

First edition 2008

British Library Cataloguing in Publication Data
A catalogue record for this book is available from the British Library

Library of Congress Cataloguing in Publication Data
A catalogue record for this book is available from the Library of Congress

ISBN: 978-0-12-373869-1

For information on all Academic Press publications
visit our web site at http://books.elsevier.com

Printed and bound in Great Britain

07 08 09 10 11 10 9 8 7 6 5 4 3 2 1

Contents

PREFACE

What is career success? The answer is simple – it is a matter of satisfaction. When an employee is satisfied with his or her career and an employer is satisfied with the employee, then the career can be deemed a success.

During one's time spent studying in school, college or university, the academic environment becomes very familiar. Many topics must be learned and good grades must be obtained consistently to be successful in school. But what does it take to become successful in industry? In addition to academic skills, specific industry skills must be mastered. Success in industry depends on an ability to comprehend and appropriately respond to industry needs.

Holding a job in industry is both exciting and daunting because the industry environment is so different from that of school. Most people will spend a large amount of time adjusting to the new situation, with its many and complex demands. Sometimes this adjustment can take months, sometimes years.

> There is **much more** to life
> than what is learned in school

Having entered industry with an academic mindset myself, I have been surprised by the number of additional important topics and behaviors essential for success, which I have encountered in industry. It seems like these are well-kept "secrets" – knowledge that is usually accumulated individually through many years of work in industry – and that knowing about these would increase employees' industry-preparedness and well-being on the job. Thus I decided to compile and reveal the "secrets" I have learned.

Perhaps the reason you wish to become privy to these "secrets" is because you are in one of many relevant situations:

1. You are soon graduating.
2. You are trying to select an academic career or a career in industry.
3. You want to secure a job in industry and want to prepare yourself.
4. You are a new employee in a company and want to succeed.
5. You are an employee in a company and you have had difficulty adjusting to your environment.
6. You are an employee in a company and you want to grow and manage your career wisely.
7. You are a hiring manager or a human resource manager and would like your new employees to learn about industry.
8. You teach or coach students about industry and industry skills.
9. You have always been in academia and would like to know more about life in industry.
10. You collaborate with industry professionals.
11. You are thinking about starting your own company.
12. You have a general interest in what it is like to work in industry.

The book is divided into two parts. In Part I, Career Choice Considerations and Job Pursuit, the first chapter contains a brief overview of the measures of success in academic and industry environments and how these environments depend on each other. Chapter 2 describes different work styles and career matches and mismatches, while Chapter 3 provides useful facts about mindset through the job application process. Part II, Professional Conduct on the Job, comprises fifteen individual chapters presenting important industry skill sets. Both technical and managerial skills are included. The majority of the chapters relate to issues that encompass many different types of industries and a few chapters are dedicated to topics specific to the biotechnology and pharmaceutical industries.

Each chapter contains a brief outline of the context in which the skill set is needed and the impact this skill set has on the company, the employee and society. This is followed by a general description of the knowledge relating to the skill set, and an anecdote elucidating the need for the particular skill set. The necessary steps for success are summarized and a brief assessment plan is given. This plan makes it possible to reflect on personal situations, skill levels and potential areas of improvement. The final chapter illustrates how the skill sets in all the previous chapters must be integrated and how industry careers can be managed successfully. These chapters will provide essential knowledge in many areas, but are by no means meant to be exhaustive.

It is my hope that you will achieve an understanding of what it means to be industry-ready and to work in industry, insights that will help you jump start your or your employee's industry career or simply improve your industry interaction. Being informed and understanding the information are the first steps towards becoming a successful professional. Practice and continuous assessment of the situation are the next.

As an extra benefit, many of the skills I describe can be used to enrich interactions in the academic environment and wherever else you may go. Practice at home too. It's a win–win.

L. Borbye

Acknowledgements

I am greatly indebted to Dr. Tessa Picknett, who encouraged me to write this book.

Part I

CAREER CHOICE CONSIDERATIONS AND JOB PURSUIT

Chapter

1

COMPARISON OF ACADEMIC AND INDUSTRY ENVIRONMENTS

TWO WORLDS AND THEIR INTERDEPENDENCE

University and industry environments contain incentives for very different products and behaviors, yet these two environments are highly dependent upon each other. Students, who comprise the new workforce, are "raised" and educated at universities. Highly influenced by university professors' knowledge, work, work ethics and personalities, students are molded according to their perception of the academic role models they encounter.

> Universities are like playgrounds for the mind where new ideas can be explored freely

The formation of basic knowledge and the additional appetite for innovation are necessary for improvement in society. The implementation of an improvement is often carried out in industry through manufacturing and dissemination of a new product.

Industry is a vehicle
for the translation
of knowledge into products
with direct impact on society

New industries are created to carry out large-scale production of a product deemed capable of becoming successful in the marketplace. The viability of production is tightly correlated with the success of the product. Therefore, it is imperative for the producer that correct assumptions are made regarding the product and its abilities, the market and the cost of production. In addition, an effective production unit must be established that complies with regulatory requirements and human resource guidelines. The correct decisions are crucial for the financial success and lifespan of the producer.

It is clear that the academic and industry environments interact and complement each other in many ways. Students need the basic education they can obtain only at the universities. Ideas and innovations are fostered in both environments and exchange takes place through publications, patents, employees, collaborations, etc. Large-scale production and commercialization are carried out in industry. A complex interchange of resources takes place. In order to fully appreciate this, it is essential to understand the two environments and their cultures in terms of their incentives, individual foci and goals.

Academia: Culture and Mindset

In the traditional academic environment, students are taught that grades are a measure of success. Grades are typically obtained by achieving average test results on multiple written and/or oral exams performed throughout the time of study. Student success, therefore, is an individual endeavor. Because national grade averages are observed, this success is likely to depend on the level of success achieved by fellow students.

Classes are often taught as traditional classroom lectures. This education method fosters a learning style where students listen to experienced masters of the "field." The intent is to let knowledge transmigrate the fastest way possible from teacher to student by immediately presenting information. Learning outcomes depend on the individual's ability to listen, take notes, understand and pass tests.

Professors also run an individual race for success, called tenure-track. To be successful, comparisons of scholarship to current and former professors are made at certain time-intervals. Instead of grades, the success of a tenure-track professor is most often measured by the amount of grant funding obtained and number of peer-reviewed publications produced while at the university. Other criteria for success are also possible.

At universities, fame is considered a virtue

For both students and professors, the university is a competitive environment, which focuses on the individual and rewards very tangible measures of success such as grades, grants and publications. One may conclude that fame is considered a virtue. Monetary compensation is usually humble and justified by the freedom to operate individual schedules and to choose research and deadlines. This ensures that only people with a desire to become researchers and/or hold so-called "secure" jobs (when tenured) enter and stay at the university as professors.

SKILLS FOR SUCCESS IN THE TRADITIONAL ACADEMIC ENVIRONMENT

Students must pass individual exams with high grades and multiple tests with favorable results. Some students are affected by the stress of intense test moments and the results they obtain do not necessarily reflect their skill sets. For such students it is necessary to know how to manage their stress. As a rule, a good study discipline, rest, exercise and appropriate nutrition are essential elements of such management. The focus on grades nurtures an environment of individual achievement. Students are conditioned to compete and protect their own interests. Most often, they are not encouraged to share or seek teamwork.

Both students and professors are conditioned to compete and protect their own interests

Professors must excel in many disciplines to succeed in the academic environment:

1. Professors must publish peer-reviewed papers based on research. A publication requires research data. Professors must be able to perform research and collect and analyze relevant data from this research. It usually takes time (years) before enough publishable data is gathered. This is the reason many universities provide new professors with an initial amount of money, called start-up funds. Professors use the start-up money to furnish their laboratories and produce enough data to publish and start applying for continuous grant funding.
2. Professors usually must obtain grant funding for their research. Grant writing requires attention to detail; the correct funding opportunities must be identified, the grant guidelines must be understood, and the grant must be written as specified and be in compliance with university guidelines. It is a long and sometimes complicated process. Professors often write several grant applications at the same time and it is expected that they "feed" a grant application pipeline throughout their career. Grant funding usually requires collection and analysis of prior data.
3. Professors must be able to supervise students and give lectures. Teaching is a joy for some professors, but for many it is considered a necessary service required by the university, a service that sometimes hampers research because of its time consumption. Student supervision, mentoring and participation on graduate committees are also required, so most professors depend on students to produce results in their laboratories.

Industry: Culture and Mindset

The focus in the industry environment is on corporate goals. These goals are typically rather complex and reaching them requires the interaction of many people. An executive team develops a business plan, which is executed by management teams and includes experts in finance, business development, intellectual property (IP), research, management and many other fields. Among the measures of success is the ability to meet deliverable demands. Some examples may include: The research group must invent new methods for investigating a topic, the intellectual property group must submit patent applications and obtain patents, or the project management group must ensure all deliverables are delivered on time. These measures of success are usually described in metrics, that is, a certain number of tangible elements, which need to be obtained by a certain time-point.

Time is a resource, a commodity and an important element in all planning. The most expensive resource is often employee time (in terms of compensation). Therefore, industry culture tends to be less flexible than academic culture concerning punctuality and effectiveness of communication.

In industry, time is a commodity

Good teamwork, interpersonal skills and communication must exist among the parties involved in reaching goals. Special achievement, like new inventions and increased responsibilities, often results in increased monetary compensation. Results are rewarded and "fame" is merely the property of the company rather than the individual.

Skills for success in the industry environment

To many people, it is surprising that so many skills are needed in industry in addition to those learned in the academic environment. These additional skills are described in detail in Part II and a brief overview is given here. Most importantly, a re-conditioning of mindset is essential.

Business goals and bottom line

Working in industry is about understanding and achieving the goals of the business and aligning personal and corporate goals. This can be achieved by knowing and respecting the company goals and by setting personal goals accordingly.

Success in industry requires that corporate and personal goals are in alignment

Leadership and teamwork

Some of the hardest things for people to change are early learning and habits. Having been taught individualism for so long, most people find it difficult to lead teams and perform in teams. Nevertheless, good leadership and teamwork are essential in industry and can be learned and perfected through training and practice.

Teamwork is a **must**

Communication skills

Many companies train their employees in interpersonal and communication skills, necessary for successful team and inter-team conduct. Communicating in an effective manner to executives, peers and subordinates, knowing how to answer questions and which questions to answer requires understanding of various target audiences. Because communication involves many different variables, one must always have both the current circumstances and the finished product in mind.

Marketing

Being able to market one's ideas and results is part of the professional communication skill set. It is important to know the various ways this can be done and that the type of language necessary depends on the kind of marketing that is performed. It also depends on where the marketing takes place, whether it is inside the company or in a public domain. Some examples of venues for marketing are staff meetings, press releases, annual reports, at public meetings and during networking events.

Discipline

What seems striking to many is the discipline that is displayed within industrial settings. People are expected to be punctual, to follow schedules and regulations, and to produce and document their deliverables on time. Regularity in work patterns is a subject for optimization and companies always work to optimize. Another area where discipline is in high regard is personal behavior. The better the individual behavior, the more positive the environment created, leading to greater job satisfaction among employees. This statement is valid, but not necessarily reinforced, in all environments, including academia.

Satisfaction on the job depends largely on personal behavior

Flexibility

As time passes, change happens. How does one cope with change? The industry environment is highly dynamic and changes should be expected at any time. Changes can relate to anything, including new decisions made by management concerning growth, the strategy, the influence of market forces, changes in the industry as a whole and implementation of new technology. Change can be a means of reaching company goals and improvements. Many companies offer training in change management to help employees adapt to phases demanding a high level of flexibility.

Flexibility is a virtue

Creativity and out-of-the-box thinking

Innovative thinking, which leads to any kind of intellectual property development, is a life-line for a company, just like grants are for a university. Such thinking can be nurtured by access to inter- and multi-disciplinary knowledge, brain storming sessions and interaction between business and finance experts and researchers. Many companies also send their employees to professional meetings and to collaborate with universities.

Ambiguity management

Most people like to make decisions they are confident are right. In companies, decisions often have to be made under constraints. Such constraints could include lack of time, data and/or information. It can be uncomfortable and stressful to make decisions like this and there are risks involved in making a decision that does not seem 100% informed.

Risk-taking is necessary

Intellectual property

Making an invention is only an asset to a company if the value of the invention is consolidated as intellectual property. Companies often employ a number of people to oversee and manage their intellectual property portfolio. The assets therein can affect how stakeholders evaluate the company in terms of commercial value and interest in collaboration, merger or acquisition.

Specialty technology and knowledge

Companies must ensure that their employees stay up-to-date with new technology and required refresher courses to stay competitive. There are many fields requiring specialty knowledge and license renewal. Examples of these fields in the biotechnology and pharmaceutical industries are genomics, drug discovery and drug production.

Quality

Pharmaceutical industries must comply with specific regulations regarding current good practices in laboratories, during clinical trials and manufacturing. These practices help ensure that the finished product is of an acceptable quality.

Ethics

Ethical conduct is of importance in all areas of work and personal behavior. There are specific regulations, with which compliance is necessary when working with human specimens, and other regulations for gene-modified organisms. These regulations are not yet universal worldwide. Often the "rules" for personal conduct remain unspoken and are a result of the interaction between people.

Globalization

Due to the trend of globalization, many companies have sister branches in other countries. It can be a challenge to make policies and regulations universal, even within one company. The major reasons for this are that people around the world have different upbringings, value systems and communication styles. Input from other cultures is an invaluable resource for innovative thinking and inspiration. At the same time, all branches of a company need to be successful, so an element of competition is likely to enter unless different deliverables are clearly assigned to the different branches.

> Cross-cultural interaction
> creates a foundation for innovation

Expectation management

For company employees, there is no such goal as tenure. Employment situations are dynamic and likely to change sooner or later. Employees are typically reviewed once or twice per year. During the review it is usually possible to get an idea of what kind of value the company assigns to the work an employee is doing. If the employee is meeting or exceeding the specified deliverables, it is likely that the employee will continue to be employed, get a raise or maybe even a promotion.

> Expect the unexpected, such as some day needing a new job

Career management

Employees are most effective when they are satisfied with their jobs. Therefore, employers want their employees to be aware of their current status and to be or become satisfied. This process requires knowledge of whether or not personal and company goals coincide. An employee can improve his or her situation as needed either by adjusting to the current job, adjusting the job in collaboration with a supervisor or by finding another job.

TRENDS OF MERGER

Academic and industry requirements are likely to develop some common trends. Whereas research and supervision are entities known in both worlds, each of the parties is now also involved in seeking public funding for enterprises and in managing intellectual property.

Universities are becoming increasingly savvy concerning intellectual property. There is a trend towards protecting academic knowledge by obtaining patents prior to publishing. Professors are requested to disclose their inventions to intellectual property and technology transfer units at the universities, and these units work to protect the intellectual property and negotiate rights and conditions on behalf of their academic staff.

Companies have realized that some of their research can be supported by grants, although this kind of funding most often is considered temporary and not market-driven. Employees can expect to be writing grant applications regardless of their choice of environment, a scenario that was unlikely not too long ago. Figure 1.1 illustrates a trend in overlap in responsibilities between jobs in academia and industry.

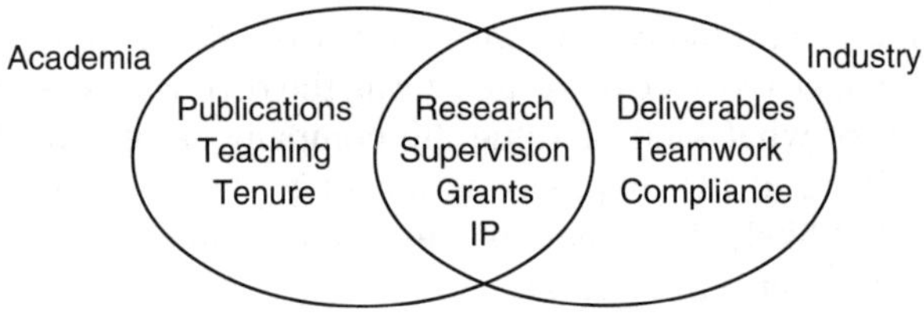

Figure 1.1 Responsibilities in jobs in academia and industry. Schematic presentation of responsibilities in the academic and industry work environments. IP = intellectual property.

Another sign of the merger of interests is the increasing number of collaborations that take place between university and company researchers. There is an advantage for industry to have ties with the academic world and its wealth of expertise. Likewise, there is an advantage for academia to have ties with industry and its access to new and often expensive forefront technology.

Identifying the compatible environment

Before beginning a career path it is helpful to be aware of some general trends in the academic and industry environments. These, of course, constitute a simplified picture of reality because the individual environments within both academia and industry vary. Table 1.1 lists some of the most common considerations.

Who a person is and how he or she likes to work determines whether or not the person can thrive in one or the other environment. If the person would rather do individual research, stay in the same environment with a minimal amount of change (also salary-wise), have a lot of freedom to operate, teach and supervise students, and secure a tenured position, the academic

Table 1.1 Comparison of the academic and industry work environments

Topic	Academia	Industry
Core unit	Individual	Team, company
Typical work style	Individualistic	Collaborative
Flexibility	Some importance	High importance
Discipline	Some importance	High importance
People skills	Some importance	High importance
Deliverables	Grants, publications	Various metrics
Tenure	Possible	Not possible
Promotion	Tenure guidelines	Depends on effort
Salary and raises	Lower than industry	Higher than academia
Special benefits	None	Stock options

environment may be the best option. If the person is people-oriented, enjoys teamwork, wants effort-driven compensation and is less interested in personal visibility, industry will most likely offer the best options for job satisfaction.

Summary of key skills for success

To be successful in academia you must be able to:

1. Obtain grant funding
2. Perform research and manage a laboratory
3. Teach and supervise students
4. Publish results in peer-reviewed journals
5. Comply with university guidelines
6. Have interest in personal visibility

To be successful in industry you must be able to:

1. Meet deliverables
2. Work well with others
3. Manage research, people and change
4. Protect intellectual property
5. Comply with company and regulatory requirements
6. Be punctual
7. Have interest in the visibility of the company

Mindset assessment: Choosing an academic or industry career

The following set of questions is useful for the assessment of your mindset to determine whether it is predominantly academic- or industry-oriented. If most of the answers are found in the Mindset A category, it indicates that an academic position is likely to be the best match. If Mindset B is predominant, it indicates that an industry position should be considered.

Question	Mindset A	Mindset B
What is your preferred work style?	I prefer to work by myself	I enjoy working with others
Are you looking forward to a tenured position?	Yes	I do not strive for tenure
How important is a high salary and effort-driven compensation?	Not so important	Very important
Do you enjoy teaching and supervision?	1. Yes 2. No	1. Yes 2. No
Do you prefer an environment of high- or low-change dynamics?	Low	Indifferent
Do you adjust to change easily?	No	Yes
Do you like to decide your own work schedule and be your own boss?	Yes	Indifferent
Can you thrive in a high-risk environment and make decisions under constraints?	I prefer not to	Yes
Is an executive demand for a change or adjustment in project(s) welcome?	No	It depends, but I will comply with the executive decision

(*Continued*)

Question	Mindset A	Mindset B
Is an executive demand for a career change or adjustment welcome?	No	1. No, but I can/will cope 2. Yes, but demanding

The two mindsets differ in several, but not all characteristics. Most professors thrive by working by themselves, whereas some focus on collaborations, teaching and interacting with their students. In industry there is also a mix of work styles and jobs. It is helpful to understand what your work style is like before you apply for a job, go to interview for a job and when you hold a job. Some examples of how work styles and careers can be matches or mismatches, and the consequences thereof, are described in Chapter 2.

Chapter

2

CHOOSING AN INDUSTRY CAREER

After reviewing the requirements for success in the academic and industry environments, it is time to analyze job options. The first step towards obtaining satisfaction within a job is to know one's preferred work style. The second step is to choose a job that matches that work style in terms of environment, type of work and interaction with other people.

> Match job and work style for satisfaction

IDENTIFICATION OF PREFERRED WORK STYLE

The academic research environment is known to attract people with a predominantly individualistic work style, although there are exceptions. In industry a variety of work styles are found due to the broad range of work that is required. The focus of this book is on industry careers. Table 2.1 shows a number of work styles. Typically, individuals will need to adopt most of these styles to varying degrees depending on the job they hold.

Table 2.1 Work styles

Work style	Description
Individualistic	Prefers working alone, is responsible for own projects and success
Organized	Manages work according to planning and timelines, thoroughly documents efforts
Analytical	Pays attention to detail, employs a systematic approach to analysis and troubleshooting
Objective	Decides action based on objective observation and facts, deduces results based on facts
Ethically rigid	Adheres to rules, regulations and guidelines, maintains quality of work and personal behavior, displays strong ethics
Collaborative	Prefers working with others, is responsible for team and individual projects and success, interacts and shares information with others, is comfortable with public presentations
People-oriented	Cares about people and their feelings, anticipates impact of decisions on people, manages people and personal interactions, has good communication skills
Performance-driven	Is highly result-oriented, seeks to improve all aspects of work and personal interactions
Flexible	Copes with ambiguity and manages change, adapts quickly to new situations, is excited about change, embraces change as an opportunity for improvement
Innovative	Creates innovations and employs innovative thinking, assumes out-of-the-box opportunities, enjoys novelty, inter- and multi-disciplinary interactions, cross-boundary work attitude

Common industry careers and work styles

Nine job types are listed in terms of the specific environment (most frequent environment listed first), the type of work (most frequent type of work listed first) and a description of the essential work styles in the given environment. This means that other work styles may also be of importance. People with other work styles may find job satisfaction within these jobs, but they may have to compromise their preferred work style.

Example 1: Researchers

Sample environment	Laboratory, office, meetings
Type of work	Independent laboratory and office work, data analysis, oral and written presentations
Work styles description	Researchers usually perform independent research and must be comfortable with the **individualistic** work style. They need to pay **attention to detail** while analyzing data, and results should be deduced **objectively**. Researchers must manage projects and document data in an **organized** manner. Researchers need to be **innovative** and **ethically rigid**, the extent of which depends on the area of research.

Example 2: Group Leaders, Project Managers

Sample environment	Office, meetings, laboratory
Type of work	Planning and troubleshooting, project and people management, oral and written presentations
Work styles description	Group leaders and managers interact often with group members and therefore need to be **collaborative.** They must be **organized** when managing projects and data and **analytical** when troubleshooting. Leaders who are **performance-driven** know how to motivate people to do their best.

Example 3: Public Relations Staff

Sample environment	Public environment, executive meetings, office, phone meetings
Type of work	Marketing and public relations, sales, writing of press releases and reports
Work styles description	People working in public relations are responsible for messages sent to the public. They must be **collaborative** and have good communication

	skills and **people-orientation**. They need to be **organized** and detail-oriented. They also need to show **flexibility** when deadlines change or new projects appear urgently.

Example 4: Human Resource Staff

Sample environment	Office, one-on-one confidential meetings
Type of work	People management, employment and benefits management
Work styles description	People working in the human resources area manage many tasks relating to people. They must be **people-oriented** and be comfortable with both the **individualistic** and **collaborative** work styles. They need to be **organized** and display high levels of **ethical rigidity**.

Example 5: Business Development Staff

Sample environment	Meetings, office
Type of work	Analysis and planning, marketing and presentations, strategy development, contract and other negotiations
Work styles description	People in business development make important recommendations about the direction of a company. They need to be **analytical** and **objective** decision-makers, be **collaborative** and **organized** and display **innovative** thinking. **Flexibility** is also important, particularly in negotiation situations.

Example 6: Quality Assurance Staff

Sample environment	Manufacturing plant, office, meetings
Type of work	Analysis, documentation, troubleshooting, interaction with regulatory agencies, personnel training
Work styles description	Quality assurance staff must be **ethically rigid** because they oversee compliance with

	regulations. They must be comfortable with the **individualistic** work style. They must be **organized, analytical** and **objective** decision-makers. When interacting with personnel or regulatory agencies, a **collaborative** work style is necessary.

Example 7: Finance and Accounting

Sample environment	Office, meetings
Type of work	Finance management
Work styles description	People in finance and accounting make and keep budgets. They most often need an **individualistic** work style and to be **analytical** and **objective** decision-makers. They need to show some **flexibility** when resources change and at the same time be **ethically rigid** in terms of compliance with laws.

Example 8: Legal Counsel

Sample environment	Office, meetings
Type of work	Legal documentation, negotiation with collaborators, legal authorities, others
Work styles description	The legal staff must be **ethically rigid** because they work with regards to the law. They must be comfortable with both the **individualistic** and **organized** work styles. The staff must display **people-orientation** when discussing and negotiating intellectual property. Decision-making must be performed in an **analytical** and **objective** manner.

Example 9: Entrepreneurs, Chief Executive Officers

Sample environment	Public meetings, confidential meetings, office, phone meetings
Type of work	Innovation, fundraising, coordination, public relations, sales and marketing

Work styles description	Being leaders of their companies, chief executive officers must set missions and goals and therefore be **innovative** and **performance-driven**. In their jobs they interact with many people and need to be comfortable with the **collaborative** work style. **Flexibility** is needed both towards change and with all stakeholders. Good leaders are often **people-oriented**, humble and **ethically rigid**.

Career Match and Career Mismatch

When an employee's career matches the employee's preferred work style, there is a high likelihood that he or she can attain satisfaction. On the contrary, if an employee is working under conditions that do not match the preferred work style, this individual is likely to experience some difficulty. Below are two examples of matched and mismatched careers and work styles.

Match–mismatch 1: The Researcher

Match: An individual researcher who enjoys working alone in the laboratory, who likes a focus on innovation, who prefers to make decisions based on logical analysis of data and who is flexible in the event that new experiments cause a change of plans.

Mismatch: An individual researcher with a very different work style, such as a highly collaborative and people-oriented individual. This individual will most likely miss out on energy derived from personal contact if working alone in a laboratory for the entire day. This person may need a large amount of personal interaction away from the job.

Match–mismatch 2: The Chief Executive Officer, CEO

Match: A CEO who has a collaborative, people-oriented work style derives joy and energy from having contact with other people. If the CEO is entrepreneurial and a factual decision-maker, innovation is spurred from creative thinking and the simultaneous analysis of data, as are many of the business decisions. Furthermore, if this individual is flexible, he or she may be comfortable exploring many options.

Mismatch: Because interaction with other people and flexibility are normally essential work style components for CEOs, the obvious mismatch is an individualistic and highly rigid work style – the person who enjoys working alone with concrete data and strict adherence to conventions. Being a CEO may be comfortable enough in a one-man company, but responsibility for a large company could cause fatigue and stress due to the mismatched work style. This individual may need time alone and stress-relief away from the job.

Overcoming career mismatch

There are two ways to overcome mismatch between work style and job. Once an individual is aware of his or her work style *and* the fact that the job is a mismatch she or he can actively change the situation – either tweak the job towards the preferred work style or change career.

Option 1: Make an effort to tweak the job towards the preferred work style

In the case of the "mismatched" researcher, this person may have the possibility to start research projects with other people, to teach science to non-science professionals, to assist marketing directors and executives with oral and written presentations, etc. There are always options to add a component such as interaction and if there is no time to do so during regular hours, lunch time may provide a good opportunity to start a seminar group or social club.

> Adding components to a job is easier than excluding them

The case of the "mismatched" CEO is less simple. This CEO prefers to have less of a single specific component strongly correlated with the job, namely interaction with other people. One way to meet this need could be to hire an assistant with a collaborative work style to manage some components of the interactions. This type of CEO would most likely benefit from frequent "big picture" overviews from directors. Furthermore, it is important for this type of CEO to get help to stay in touch with the employees, as well as practicing flexibility.

Option 2: Change career

The "mismatched" researcher seems to have possibilities for improvement by adding more personal interaction. The "mismatched" CEO may find it more difficult to exclude personal interaction. Depending on the difficulties that are encountered, it may be wise to opt for a different career. It is beneficial to view this as an opportunity instead of as a personal failure. Many people change careers and also work styles over time.

Summary of key skills for success

To be successful when choosing an industry career you need to:

1. Know your work style
2. Know which careers match your work style
3. Understand how to tweak your job towards your work style
4. Take a new job if your current job cannot match your work style

Mindset assessment: Knowing your preferred work style

Are you knowledgeable about your own work style and is your job a match with it? The questionnaire below will help you determine this. If your answer to most questions is "Yes" you are most likely on the right track. If your answers are "No" these may be areas in which you can improve your situation and understanding (or, of course, the question may not apply to you).

1. Do you know your work style?
2. Do other people agree with your view of yourself?
3. Are you able to identify jobs that match your work style?
4. Are you able to apply your work style to your current job?
   a. If Yes, stop here.
   b. If No, are you able to tweak your job towards your work style?
      i. If Yes, stop here
      ii. If No, are you ready to look for a new job?
5. Do you want to be a researcher?
   a. If Yes:
      i. Are you comfortable with the individualistic work style?
      ii. Do you pay attention to detail?

        iii. Do you deduce results objectively?
        iv. Do you document data in an organized manner?
        v. Do you like to invent?
        vi. Are you ethically rigid?
    b. If No, continue below
6. Do you want to lead a group?
    a. If Yes:
        i. Is your work style collaborative?
        ii. Are you organized when managing projects and data?
        iii. Are you analytical and good at troubleshooting?
        iv. Are you performance driven?
    b. If No, continue below
7. Are you interested in public relations work?
    a. If Yes:
        i. Is your work style collaborative?
        ii. Do you have good communication skills?
        iii. Are you organized and detail-oriented?
        iv. Are you comfortable with a high level of flexibility?
    b. If No, continue below
8. Do you want to work with human resources?
    a. If Yes:
        i. Are you people-oriented?
        ii. Do you work well with both the individualistic and collaborative work styles?
        iii. Are you organized?
        iv. Do you consider yourself ethically rigid?
    b. If No, continue below
9. Are you interested in business development?
    a. If Yes:
        i. Are you analytical and objective in your decision-making?
        ii. Is your work style collaborative and organized?
        iii. Do you think innovatively?
        iv. Are you flexible when you negotiate with people?
    b. If No, continue below
10. Do you want to work in quality assurance?
    a. If Yes:
        i. Do you display a high level of ethical rigidity?
        ii. Do you perform well in an individualistic work style environment?
        iii. Are you organized, analytical and objective?

    iv. Are you comfortable using a collaborative work style when necessary?
  b. If No, continue below
11. Do you want to work in finance or accounting?
  a. If Yes:
    i. Do you display an individualistic work style?
    ii. Are you an analytical and objective decision-maker?
    iii. Are you both flexible and ethically rigid?
  b. If No, continue below
12. Are you interested in being on the legal staff?
  a. If Yes:
    i. Are you ethically rigid?
    ii. Can you work well in an individualistic work style environment?
    iii. Are you organized?
    iv. Are you people-oriented?
    v. Are you analytical and objective?
  b. If No, continue below
13. Do you want to start your own company?
  a. If Yes:
    i. Are you innovative?
    ii. Are you able to motivate people to do their best?
    iii. Is your work style collaborative?
    iv. Do you like to be flexible in most matters?
    v. Are you humble and people-oriented?
    vi. Do you display a high level of ethical rigidity?
  b. If No, there may be other opportunities for you.

Knowing your work style is a key factor in the identification of a suitable career. Once you have identified a matching job, the job offer must be pursued and obtained. Chapter 3 describes this process in detail.

Chapter

3

PROFESSIONAL CONDUCT DURING THE JOB APPLICATION PROCESS

The reasons a person may be looking for a job include a change in degree status (graduation), unemployment, or a wish for changes and new opportunities better matching goals and work style.

Getting a job offer and securing a job is a process with many steps. These steps include: Choosing the job for which to apply, applying for the job, interviewing for the job, giving feedback after the interview and final negotiations. This chapter describes the goal, the specific elements and the needed awareness necessary during each of these steps.

JOB DESCRIPTION ANALYSIS

Goal: Find a job that matches personal priorities.
Mindset: Think about what is most important and prioritize.

The first step is to determine what is of highest priority. Typically, the following elements are deemed relevant:

- Location
- Educational level

- Full-time or part-time job
- Salary
- Benefits
- Specific skills (technical and personal)
- Type of work
- Work style
- Organizational structure.

> Focus on your most important needs when you look for a new job

Priorities vary and depend on personal preferences. For example, if being together with family is most important, then location is probably on the top of the list. If money is the main concern, then salary will be of top priority. If working with other people is essential, there must be opportunities to interact. Contrarily, if the individualistic work style is preferred, there should be time to work alone.

The analysis of a job description can quickly provide information for a decision concerning whether or not to pursue the job. Acquiring additional information may be necessary.

Example job description: Marketing and Sales Manager

We are seeking an independent and people-oriented individual with good interpersonal and communication skills to assist with our medical sales and marketing operation world-wide. Together with the Director of Marketing and Sales in our New Jersey location, you will help oversee our initiatives in our locations in Europe and China.

You have experience from a similar position, a Master's degree in Science and a minimum of two years of corporate training. Language proficiency in Mandarin, German and French as well as fluency in English is preferred. You are an experienced MS-Office user. Some experience with cross-cultural management is an advantage.

We are offering an exciting international job in a growing company with room for professional and personal development. Salary and benefits will be commensurate with experience. The job is a one year trial basis appointment with possibility for extension up to three years. You can read more about our company at . . . or contact . . . .

Example job description analysis: Marketing and Sales Manager

Location:	New Jersey, Europe and China
Educational level:	Master's in Science and two years' corporate experience
Full-time or part-time job:	Full-time, trial basis one year, maximum three years
Salary:	Commensurate with experience, need more information
Benefits:	Commensurate with experience, need more information
Specific skills:	Mandarin, German, French, English, MS-Office, interpersonal skills, communication skills, cross-cultural management, travel
Type of work:	Assisting Director (?), need more information
Work style:	Independent (individualistic work style), work with Director, work with teams in Europe, China and New Jersey (collaborative work style)
Organizational structure:	Reporting to Director, need more information

Often it is a good idea to call the contact person to find out exactly what is expected of the applicant and if there are internal candidates applying for the position.

THE JOB APPLICATION

Goal:	Obtain an invitation for an interview.
Mindset:	Think about what the employer needs and address those needs.

After a job description that matches the personal priorities and preferred work style has been found, an application for the job needs to be written. Sometimes the application submission is required on-line and sometimes a hard copy is desired. The on-line submission procedure may specify a certain format and allow for the enclosure of multiple attachments. A hard copy version usually allows more freedom to decide the format. An application typically consists of the following components:

1. Cover letter
2. Curriculum vitae (resumé)
3. References
4. Other relevant documents.

How the documents are presented (packaged), and whether or not the application deadline is respected, will contribute to the likelihood of receiving an invitation to an interview.

Cover letter

A cover letter should contain the following components:

- The applicant's contact information
- Where the job description was found
- The applicant's professional background (emphasizing the employer's needs and what the applicant will bring to the position)
- The applicant's personal skills (again emphasizing the employer's needs)
- A closing phrase about how this job is a great opportunity, that further information can be provided and that a response from the employer is anticipated
- The contact information for references
- List of enclosures (if hard copy).

Focus on the employer's needs

Three common mistakes are:

1. Any kind of negativity
2. A wish to give impressive information that is not true or not relevant for the position
3. A personal judgment praising one's own capabilities ("I will be a great addition to your company").

The phrasing needs to be concrete, to the point, positive, enthusiastic, truthful so it gives a realistic picture of the applicant and addresses the exact needs of the employer.

Always tell the truth

The cover letter should not exceed one page.

Curriculum vitae (resumé)

The curriculum vitae (cv) is a personal "professional life document." It contains:

1. The applicant's contact information
2. A list of degrees from educational institutions
3. Professional experience, including teaching experience and consultancies
4. *Pro bono* work
5. Publications, invited seminars, patents, grants, etc.

The length of the cv depends on the level of the position applied for. For example, an application for an entry-level position may be accompanied by a 1–2 page cv, whereas an application for a senior scientist position may justify 5–10 pages because of the applicant's prior relevant experience. People with academic experience may also include a list of students and graduates and other items such as workshops and advisory boards.

The style of the cv should be **concise** and individual items should be easy to find. All items should be listed in order so that the most current information is listed near the top. Use headings and do not use extra words or long descriptions. A good cv should be able to accompany any job application after minor modification.

Details that might be given on the cv

Contact information:	Applicant's name, address, phone number, e-mail
Educational background:	Degree, institution, address, year
Professional experience:	Job title, company, year, responsibility
Pro bono work:	Job title, company, year, responsibility
Publications:	Name(s), title, journal, page and year
Seminars:	Name, title, meeting, location, date, year
Patents:	Patent publication location, serial number, title, date, year
Grants:	Granting agency, funds, project title
Other:	Personalize as necessary

Letters of recommendation

Obtaining strong letters of recommendation is essential for chances to gain employment.

> In letters of recommendation potential employers look for signs of negativity or omission of good traits

Confidential letters of recommendation are usually submitted directly to the employer. Because they are confidential, they are highly valued. Non-confidential letters of recommendation are also valuable, but because these may have been seen by the applicant and discussed with the person(s) providing the letters, they may be deemed as less important. Different countries and companies have different rules regarding letters of recommendation. Sometimes the letters must be submitted together with the application (in which case the materials have been seen by the applicant), sometimes they are submitted directly to the companies by the persons providing the letters of recommendation, and in other cases the referees themselves may be contacted directly by the companies.

People who provide letters of recommendation should address the applicant's acumen in areas of specific interest to the employer and the specific position. They should also address the applicant's interpersonal skills. Finally, they should be willing to be contacted by the employer if more questions arise after reading of the reference letters or later in the process of further interaction.

Applicants must select those who provide references with great care. The people who are writing about the applicant must be trusted to write positive statements. If there is any doubt about whether a person is going to provide a good recommendation, it is wise to find somebody else to write it.

> Pick trustworthy people to provide references

Other relevant documents

Any documents that can be used as evidence of experience relevant for the position can be included. Such evidence could be a syllabus from a course, a workshop, *pro bono* work descriptions, flyers, etc.

Include relevant documents only

Packaging

The entire application should be checked for spelling errors, style and format, and look as professional as possible. For this purpose clean white paper and fonts that are easy to read work best. The application package should be organized in a logical manner with the cover letter on top followed by the cv and other supporting documents.

Make the application look professional

Deadline

In order to be considered for an interview, the application must be received by the employer "on time." "On time" means if there is a deadline, the applicant must make sure the application is at the employer before or at that time. If there is no deadline, it is advisable to send the application as soon as possible because the job will most likely be filled as soon as an ideal candidate is identified.

Respect the deadline

INTERVIEW PREPARATION

Goal: Become confident and well-prepared for the interview.
Mindset: Prepare both mind and body to do their best.

Research company goals

Before going to the interview it is wise to have a basic understanding about the company goals and culture. Simple questions like these can be researched in advance:

1. What does the company produce?
2. What kind of technology is employed?

3. How big is the company (small, medium, or large)?
4. Where is the company located (one or many locations)?
5. Is it a publicly traded company?
6. What is the financial situation (stock price if publicly traded)?
7. Which partners and collaborators exist?
8. What do the most recent press releases describe?

Most of this information can usually be retrieved from the company website. If the area where the company is located has a central industry information center, it may be worthwhile to obtain information locally as well.

Practice communication

Often the hardest part of the conversation is the first five minutes. Much stress can be alleviated by practicing what to say when meeting with people for the first time. Thanking them for the invitation is a very good way to start an interview. An introduction using a few sentences relevant to the job is a good way to continue. Also taking the necessary time to let the interviewer(s) introduce themselves is good practice. This attitude shows politeness, patience and interest in people.

Practice being grateful for the interview

There are certain questions that can be anticipated and an applicant can practice answers to these in advance. Some of the most common questions can be found on the Internet and typically relate to the following topics:

1. Experience and education (capability)
2. Five-year goals (assessment of ambition)
3. Work style (priority of components)
4. Worst experience (coping skills)
5. Strength and weaknesses (assessment of self-evaluation)
6. Flexibility (coping skills)
7. Why this job seems interesting (assessing reasons for application).

Practice answering questions

One of the most common mistakes during an interview is to be negative. When applicants are asked to describe their worst experience (a common request) they can choose something relevant to the new job and focus on how challenging the experience was and how well it was handled. Likewise, when describing weaknesses, a weakness that may also be considered a strength, can be chosen.

> Negativity is a common mistake – avoid it

Eat well

Taking the time to nurture the body will help with concentration and mental balance. It is a good idea to eat well (healthy food) the day and morning before the interview.

Exercise

Physical activity is good for both the mind and the body. It is optimal to exercise the day before the interview. Another kind of exercise that may give a competitive advantage is mental imagery, especially if nervousness is a prevalent problem in interview situations. A person can imagine him or herself successfully navigating a job interview. The technique involves focusing on seeing the event as a movie and playing it several times so the scenario is easy to remember.

> The use of imagery can improve confidence

Rest

A greatly underappreciated part of interview preparation is sufficient rest. An applicant should avoid being physically or mentally exhausted to display the best performance. If traveling is involved to get to an interview, it is wise to arrive early so there is time to adjust and rest.

Dress well

Obviously good hygiene and a professional look are important. There is also usually a corporate dress code. Information about the dress code can be requested from the company's HR Department or these simple conservative rules can be followed:

Women:

- No open-toed shoes
- Heels maximum 1.5 inches (3 cm)
- No tight or revealing clothing or mini skirts
- As little make-up as possible
- No or very little perfume
- No excessive jewelry
- Classic attire (jacket and pants, jacket and skirt).

Men:

- Classic attire (white shirt, tie, suit, classic shoes).

The interview

Goal: Obtain the offer letter.
Mindset: Obtain the offer letter; evaluate the offer only *after* the job has been offered.

If the hiring manager determines that the applicant's professional background is a match with the position, the applicant will be invited to an interview. During this interview the hiring manager will evaluate the applicant's truthfulness and personality. Most people think they have to know whether or not they want the job during the process of getting the offer. However, only *after* the offer is given is it time to evaluate all the information that has been gathered and to assess whether or not to continue pursuing the job. Following this approach ensures a focus on obtaining the job offer.

Interview settings

There are many ways to conduct an interview. The applicant may meet with one or several people and may meet them one at a time or in a group. The interview may take thirty minutes or a whole day. If the interview includes a presentation, this part may be performed for a very large group before or after the interview with a smaller group. Interviews can also be performed by video conferencing or conference calls.

The first five minutes

The beginning of an interview determines how it will proceed. This is why practicing meeting with the hiring manager and thanking her/him for the invitation to come is a good idea. Being relaxed, positive, interested, polite, kind and remembering to smile are all essential, as are posture and the signals sent as body language. Interrupting the interviewer should be avoided and active listening (see Part II, Chapter 6) practiced. When prompted for questions these should be asked in a humble manner. When it is time to depart it is essential to thank everybody for his or her time and interest.

Communication styles

During the interview a number of different communication styles and questions may be experienced. This variety may be carefully and purposefully planned to assess the applicant's reactions, or may simply be a reflection of the current environment within the company. Some interviewers may want to make the applicant feel comfortable while others have the opposite agenda. It is not uncommon for an interviewer to ask very challenging questions to test whether the applicant will say something negative, and show a range of emotions. It is essential to stay calm and objective during such circumstances.

> Stay calm when under pressure

Depending on the company location (country) there are rules for how an interview can be conducted and what can be asked. It is wise to know in advance which questions are considered appropriate (legal) and inappropriate (illegal) to pose. Illegal questions (in the USA) often relate to personal situations or resident status. If a situation is encountered where an illegal questions is posed there are several options. The applicant may choose to tell the interviewer that the preference is to not answer, that the asked questions addresses a personal issue and is considered private or simply answer, knowing that this questions should not have been asked.

Common pitfalls

Some common pitfalls encountered in job interviews and the typical perception of these "red flags" by hiring managers are listed in Table 3.1.

Table 3.1 Job interview pitfalls

"Red flags"	Possible perception by hiring manager
Negative reflection (speaking negatively about previous job, boss, salary, etc.)	Complaining, not loyal
Excessive emotion	Not professional
Excessive self-promotion	Big ego, no interest in others, too self-reliant
Interrupting interviewer	Lack of good communication skills
Rambling	Lack of overview and focus
Lack of questions	Too quiet, seems unprepared, uninterested
No eye contact	Too shy, insecure, hiding something
Restless	Uncomfortable, nervous
Monotonous voice	Boring, uninterested
No smile, facial expression	Negative, uninterested
Poor explanation of experience	Poor presentation skills, poor technical skills, poor marketing skills
Inappropriate dress code	Not professional

THE FOLLOW-UP

Goal: Show continuous interest in the job.
Mindset: Receive job offer; be thankful for and re-affirm interest in the position.

After the interview the applicant can choose to continue to show interest in the job by sending a "thank you letter" to the employer and the people who conducted the interview. This serves as a follow-up message and will signal a grateful attitude and respect for the people involved in hiring. It also reminds the hiring manager about the applicant's contact information and willingness to communicate. In addition to this letter, it is wise to later follow up with a phone call to monitor the status of the hiring process (unless there are other specifications).

JOB OFFER AND NEGOTIATION

Goal: Tweak the job offer if necessary.
Mindset: Find a balance between wishes and reality.

Analysis of compensation package

When an offer letter is received, it is time to assess whether the priorities and most important aspects are met. Both the monetary compensation and environment and life style issues should be analyzed.

Elements of compensation may include:

1. Salary
2. Indirect compensation
   a. Stock options
   b. Other
3. Benefits
4. Healthcare
   a. Pension/retirement contribution
   b. Vacation
   c. Sick leave
   d. Sabbaticals
   e. Other.

When to negotiate

What is the market value of the requested job and job services? Is the market value the highest priority or is there room to compromise salary against quality of life, freedom to operate or more vacation time? Is extra time to do research or write patents necessary? Is more personal interaction necessary? These are some of the questions one may ask when reviewing an employment offer and deciding if and how to negotiate. Choosing to negotiate may pose a risk of losing the job offer or creating animosity. On the other hand, a certain level of negotiation is usually expected. It is important that both employee and employer can be happy with the agreement, therefore it is a fine balance.

Summary of key skills for success

To apply for a job you need to:

1. Know what you want
2. Select job descriptions that match your needs and background

To write a job application you need to:

1. Write a cover letter with a focus on the employer's need
2. Restrict the cover letter to one page
3. Write only truthful and concise statements
4. Avoid self-praise
5. Check your spelling
6. Trust your references
7. Provide a professional-looking curriculum vitae
8. Avoid documents irrelevant for the given position
9. Follow all instructions and respect the deadline

To prepare for a job interview you need to:

1. Investigate the company history
2. Practice answering standard questions in a positive manner
3. Practice posing questions
4. Imagine yourself being successful at the interview
5. Select and prepare professional attire
6. Get rest, good food and exercise

To interview for a job you need to:

1. Stay calm under all circumstances
2. Smile, take your time and make eye contact
3. Be polite, kind, attentive, interested
4. Use the answers and questions you prepared
5. Avoid any negativity
6. Avoid negotiations and salary discussions unless prompted

To follow up after the job interview you need to:

Send a letter to thank everybody with whom you interviewed and show interest in the position

To analyze a job offer you need to know:

1. What you need, what you want and the difference between the two
2. In which areas you can compromise
3. In which areas the company can compromise

To accept a job offer you need to know:

1. Whether you are satisfied with the offer
2. If the offer does not meet your needs and requires suggestions for improvements without definite demands

Mindset assessment: Job application, interview and negotiation

Do you know how to successfully assess whether a job is a potential match with your needs? Do you know how to successfully apply for and secure the job? The following questions will help you assess these matters. If you answer "yes" to most questions you will most likely be successful in your pursuit.

1. Searching for a job:
   a. Do you know your needs and the priority of your needs?
   b. Are you highly selective in your job search?
2. Applying for a job:
   a. Do you address all the employer's needs in your application?
   b. Do you tell the truth in your cover letter and cv?
   c. Do you trust your references?
   d. Does your application package look professional?
   e. Have you removed all irrelevant information?
   f. Are you meeting the application deadline?

3. Interviewing for a job:
    a. Have you prepared yourself properly for the interview?
    b. Do you know how to answer the most common questions?
    c. Do you know which questions to ask?
    d. Can you explain and describe all information you have provided in detail?
    e. Are you at ease and committed to remaining calm and positive during the interview?
    f. Do you know what to wear for the interview?
    g. Is your perception of yourself in line with how others see you?
    h. Can you listen well to others?
    i. Are you aware of your body language and how to improve it?
    j. Do you know how you would like to greet the interviewer?
    k. Can you detach emotionally when under pressure?
    l. Do you have good things to say about your former experiences?
    m. Are you committed to avoiding appraising yourself untruthfully?
    n. Do you make eye contact and do you smile during the interview?
    o. Do you remember to follow up after the interview?
4. Negotiating the offer:
    a. Are you ready to accept a good offer?
    b. Are you able to negotiate and be realistic at the same time?

If you have read this far you are most likely interested in an industry career. In the previous three chapters you have gained knowledge about the differences between the academic and industry environments and the overall conditioning that takes place in the two settings. Also, you have learned about work styles and how they constitute a basis for selecting an appropriate career. Using the guidelines for analyzing the contents in a job description and assessing if it is a match or a mismatch with your work style can be part of the process you use to acquire a new job. You also have knowledge about how to prepare for a job interview and how to conduct yourself during and after the interview.

Part II of the book contains fifteen chapters that reveal a wealth of essential "secrets" in industry. Some of these may seem like common sense while others may be very surprising. Taken together, all of these will provide a fuller picture of the path to success, and hopefully this will become evident and be relevant and applicable to you and your career.

Part

II

PROFESSIONAL CONDUCT ON THE JOB

Chapter

4

BUSINESS GOALS AND BOTTOM LINE

Having a plan means that somebody took time to assess goals to reach, and developed a process to reach them.

SCENARIO AND IMPACT

A company's plan is called a **business plan**. A good business plan is essential for any company.

> A company should document its direction, goal and process to reach the goal

A number of different aspects of a company are described in the business plan. These include a statement of why the company exists and for which purpose (the general mission and goals), an overview of the company composition and organization (employees and their reporting structure), a plan for the business of the company (direction of work, what is produced and when, aspects of competition and procedures for maintenance and growth).

One must learn what the **bottom line** is to fully comprehend a company's business goals. The bottom line can be described as the minimum deliverables by which the company can exist. Contribution to attaining the bottom line is the most crucial job for any company employee.

> Employees must know and contribute to the company's bottom line

In contrast, imagine working for a company that has no or no clearly defined plan. There would be confusion, questions to ask about what to do next and to whom to listen. Working without a clear plan due to a lack of goals, overview and effectiveness would likely shorten a company's lifespan. Clear plans implemented in a late phase may restore viability if the initial loss of resources is reversible.

> Lack of planning can cause lack of employment

During the job interview process and the initial phases of employment, it is wise to ensure that the company has a clear and viable business plan. The bottom line should be well-known and attainable, and the process to contribute to it should sound reasonable.

An employee must find a **place, role** and **professional identity** within the company. The sooner this happens, the more productive and satisfied the employee will be. Since these are key elements to a company's success an early and thorough definition of the expectations for all employees is advantageous. Consider some of the questions listed below:

Place

1. How and where does the work fit into the "big picture"(business plan and bottom line)?

Role

1. Who is the boss and who are the colleagues (reporting structure)?
2. What is the job (specific deliverables, contribution to bottom line)?

Professional identity

1. Are the company mission and goals attainable?
2. Are the job and the reporting structure acceptable?

IMPORTANT KNOWLEDGE

Roles and reporting structure

There are several common roles in a company. The leader of a company is often called the President or the Chief Executive Officer (CEO). The CEO may report to the Board of Directors if such a Board exists. Everybody employed within the company is responsible to the CEO. The CEO may employ a staff of Chief Officers (CXOs) or Vice Presidents (VPs). These officers usually report directly to the CEO.

Some common Chief Officer roles

- CEO: Chief Executive Officer/President
  Responsible for company's business plan and actions, and overall company success
- CFO: Chief Financial Officer/VP of Finance
  Responsible for company's financial plan and actions
- COO: Chief Operational Officer/VP of Operations
  Responsible for company's buildings, logistics and infrastructure
- CSO: Chief Scientific Officer/VP of Research
  Responsible for company's research strategy and projects

Some additional offices with common Vice President representation

- Business Development: VP of Business Development
  Responsible for product strategy and market analysis
- Legal Counsel: VP/Director of Intellectual Property
  Responsible for company's intellectual property portfolio
- Human Resources: VP/Director of Human Resources
  Responsible for employment and employee satisfaction

The CXOs may employ a number of directors to perform various tasks. For example, while the CSO is responsible for all research and research strategies, he or she may need several people to oversee research in different areas of expertise. Imagine a company that performs research in a number of areas involving humans, plants and microorganisms. Most likely, the CSO needs to have directors in each of these areas. Each director employs a group of people known as the director's staff. This staff can consist of project managers, senior and junior scientists, and technicians.

The company reporting structure described above is a **top-down management** approach. There is a centralized, leading figure at every level. It is a model in which responsibilities are easy to assign. An alternative model is the **matrix organization**. In a matrix, people and groups work in a cross-hierarchical and peer-like fashion. For example, a director can interact with more than one CXO, a scientist with multiple managers and other scientists, and technicians can have deliverables to many different scientists and managers. Assignment of responsibilities in the matrix model is more complicated than in the top-down approach, but often is highly effective for a small company or smaller units of a larger company.

The business plan

There are several components of a business plan. Knowing each of these creates understanding of the entire enterprise – its basic principles, structure and plans for the future. The entire business plan may only be accessible to key employees, while certain parts may be available to a larger number of employees. There are several common components of a business plan:

1. *Executive summary:* This section precedes all other sections. It gives a brief overview of the major points and conclusions within a business plan. The summary is often short, half a page to two pages. Reading the executive summary is a quick way to learn about a company.
2. *Mission and goals:* A company's mission is a large-scale perspective describing the specific impact it wishes to have. For example, a company may have the mission to eradicate a certain disease by preventing or eliminating it. Mission statements are very short, one to three sentences. The goals are tangible products, methods or other means by which to achieve the desired impact. One goal may be to produce a new medicine targeting the disease. Just like the mission statements, goals are stated in few sentences.
3. *Employees and organizational structure:* The Board of Directors and all of the executives are usually listed in the business plan, together with their reporting structure. Depending on the size of the company and level of detail presented, brief resumés may also be presented and directors may be included. The overall structure of the company is

often illustrated using organizational charts visualizing the different groups and departments.

4. *Financial strategy:* The financial overview and strategy includes a description of current funds, revenue, burn rate and a capitalization plan. The strategy for future financial viability of the company is described. This could include venture/angel funding, initial public offering (IPO), intellectual property asset planning, new contracts and aspects such as expansions or reorganizations.
5. *Technology, products and timelines:* A company's development and use of technology is described in this section. Scientific research may be included. Specific products relating to the mission and goals statements may appear in project descriptions. Timelines for development and/or delivery of these products are usually outlined as well as the general plan for management of the projects and deliverables.
6. *Business analysis:* Understanding the market for the products on which the company focuses, the status of the competition and predicting future consumer interest are imperative for success. This section includes analysis of the market and of competing products, which are similar or identical to the products the company intends to develop. An important part of this analysis is the risk assessment plan which outlines the "threats" to success.
7. *Product development plan:* Based on the business analysis, a plan for development of the company's products is presented. This section may also describe how current products can be altered or used in a new fashion, a product line extension potential. The plan may include ideas for the marketing, launch and maintenance of these products.
8. *Intellectual property strategy:* During the course of research and development, intellectual property is likely to be produced. The company must have a plan to capture and protect it. This plan may describe the internal procedures used to consolidate intellectual property and also provide a detailed plan and budget for patent applications and maintenance.
9. *Value statements:* In addition to the above-mentioned components, other sections relating to a company's individual image, culture, rules and policies may be added.
10. *Appendices:* The final section contains relevant appendices. These may include employee resumés, organizational charts, timelines, definition of specific technology, competitor information, lists of current publications, presentations and intellectual property, a glossary and other information.

Stakeholder influence

Stakeholders are all people who have an interest in the company's success. Financial investors (including stock holders), the Board of Directors, all employees and their families, and society are all stakeholders.

Financial investors are interested in the fiscal returns on their investments. The Board of Directors is interested in the fiscal returns as well as prospects for the future such as growth, new contracts and more revenue. Employees hope to have continued employment, a pleasant work environment and good financial compensation. People who own stocks and stock options are interested in an increasing stock price. The employees' families have interest in solid performance and stability of employment. Finally, society is interested in all of the above as well as a positive impact on the local economy based on company revenue, number of jobs and the effect the products have.

> Stakeholders have a variety of interests and amount of influence

In an ideal world, all of these needs are met. In reality these are often conflicting interests and priorities must be established. Typically, stakeholders have different levels of influence. Stakeholders with the highest amount of influence are usually those who support the company financially and/or with their professional reputation.

> Stakeholder attitude determines future action

If all stakeholders are satisfied, the company is likely to follow its current course. On the contrary, if the stakeholders show dissatisfaction, a correction of course is likely underway.

Common deliverables

There are a number of considerations to make when prioritizing which is the most important deliverable for a company. New employees must learn these priorities from others and from understanding the business plan. Deliverables may be tangible (hard) and less tangible (soft).

Common "hard" deliverables
- Products
- Intellectual property
- Contracts
- Licenses
- Publications

Common "soft" deliverables
- Improvements
- Reputation
- New interactions

What the priority list looks like depends on the immediate need of the company. The company may most need products for sale, a great reputation through publications and contracts, or improvements in its protocols for production. Maybe intellectual property is most needed in order to gain a competitive advantage and ownership in a certain area. Priorities are a function of many factors and change depending on the size of the company, the market dynamics and much more.

Priorities change

Differences between small, medium and large companies

Small, medium-sized and large companies have different priorities and ways to operate. The definition of size varies. Here, a small company is defined as a company with a maximum of 50 employees. A medium-size company employs 50 to 500 employees and a large company has more than 500 employees. Table 4.1 presents some of the common characteristics adhering to variation in company size.

Employees in small companies have many and often changing roles. The situation is highly dynamic and the level of management is often low. In a medium-size company a higher order of structure is apparent. A certain level of management is in place throughout the company and employees

Table 4.1 Company size and focus

Topic	Company size		
	Small	Medium	Large
Number of employees	0–50	51–500	More than 500
Role of employee	Multiple	Few	Highly specific
Level of dynamics	High	Medium	Low
Level of management	Low	Medium	High
Company priorities	1. Technology 2. Reputation 3. Growth 4. Contracts	1. Technology 2. Production 3. Intellectual property 4. Contracts 5. Reputation	1. Production 2. Intellectual property
Executive communication	Frequent	Some	Infrequent

still have a number of roles. In large companies, both the management structure and the employee roles are highly defined.

> How many roles you will have depends on the size of the company

Small companies find it necessary to focus on technology and to obtain a reputation in their field(s). They often do not yet have a production or intellectual property pipeline established, but look for partners and contracts. The larger the company, the more the focus often shifts towards production and management of the production. These companies are very interested in other assets, including intellectual property. Their reputations are already established, so there may be less of a need to publish and make the company known.

In a small company, the CEO may know all the employees by name and interact with them on a regular basis. The larger the company, the less direct contact there often is between the top executives and the other employees. This is due to the many additional responsibilities related to an increase in company size.

Anecdote: Aligning personal and corporate goals

Peter had been working on his project for five years. He had become one of the most knowledgeable people within his field and his research was at the forefront.

Nevertheless, despite the great academic stimuli the project was providing it was clear that an economically tangible outcome was needed. Peter's manager, Mindy, had been told by the upper administration that a five-year-old project either should be profitable, self-sustainable or terminated in order to make good business sense.

Peter had focused on providing academic publications during the five years the project had been ongoing. The company's marketing team had been using Peter's results and publications through the past two years to attempt to attract interested partners, such as investors, collaborators and licensees, without luck.

Mindy went to see Peter and she told him that the project had to be terminated. The reason she gave was that the company had not been able to initiate negotiations leading to sustainable contracts within the given timeframe. Peter would have to work in another group on an already sustainable project.

Peter was very surprised. He had spent all his time working on this project and he knew it was academically sound. Was the administration not aware of this fact? He was now supposed to work in another group and he felt that this was a disadvantage because the new project was unknown territory to him. Thoughts about how it could impact his status and salary went through his head. Peter started to become a disgruntled employee.

Mindy had predicted Peter's response. It was typical for employees who had been in the same project since their hire. She decided to help Peter. She made sure he was attending a workshop about flexibility at the workplace.

During this workshop Peter learned that working in a company is ***not about oneself or one's project***. It is all about the company business and the bottom line is what matters. He learned that if the projects are not sustainable then the company is not viable. If the company is not viable then a job in the company is not either. He learned that flexibility matters and that individual egos need to be put aside. It is ***not about being the best, but rather about together making the best happen***.

Summary of key skills for success

To fit well within the company you need to know:

1. Your role in the company
2. To whom you report
3. The mission and goals for the company
4. The deliverables and their priority levels
5. The specific deliverables you have to produce
6. Who the stakeholders are
7. The size of the company and how it influences the dynamics within the company

Mindset assessment: Business goals

Do you understand your company's business goals and bottom-line, and are you in tune with what you need to do?

The following questionnaire will help you assess yourself. If you find yourself selecting the answers in column A most often, you are probably doing all you can to understand and produce what the company needs. If you select answers in column B, you can most likely improve your own attitude as well as the work you do for the company.

Question	A	B
Do you know the company's mission and goals?	Yes	No
Do you know which kinds of products your company produces?	Yes	No
Do you understand how these products are produced?	Yes	No
Do you know which role you play in the production?	Yes	No
Do you know what your deliverables are?	Yes	No

Do you know the priority of your deliverables?	Yes	No
Do you plan your work before you do it?	Yes	No
Do you know to whom you report?	Yes	No
Do you have a good professional relationship with your boss?	Yes	No
Are you in a matrix management environment?	Yes or No	Don't know
Are you in a top-down management environment?	Yes or No	Don't know
Is it important for the company that you produce publications?	Yes or No	Don't know
Are you interacting with other employees in your company outside your group?	Yes	No
Is your company stock publicly traded?	Yes or No	Don't know
Do you know who the most influential stakeholders are?	Yes	No
How big is your company?	Small, medium or large	Don't know
Do you have many roles in the company?	Yes or No	Don't know
Do you understand the difference in dynamics between small, medium-size and large companies?	Yes	No

Chapter 5

LEADERSHIP AND TEAMWORK

Leaders are found everywhere; they exist at all levels and certain people are so-called natural leaders. Being a leader is a position of high responsibility. A leader must honor this responsibility by producing appropriate results, both technically and socially. Contrary to popular belief, excellent leaders do more than lead; they serve a cause and other people.

> Good leaders lead *and* serve

SCENARIO AND IMPACT

A leader/boss is anyone who is responsible for another person's deliverable. For example, the CEO is responsible for the deliverables from all employees, a director is responsible for a group's deliverables, a manager is responsible for several teams' deliverables, and a team leader is responsible for the deliverable from the team.

Teams are the predominant work units in industry and are considered superior to the individual worker. This is because there is a belief that several people can get more done than one person, and that the work that needs to be done is complicated and requires simultaneous expertise from several individuals. Teams are often led by a team leader except in cases where a true matrix organization is preferred.

> Leaders and teams depend on each other for decisions and data

Leaders are appointed because of a belief in the higher effectiveness of a chain of command rather than distribution of equal decision-making power to all. This is especially true as a company grows. The vision, goals and the ways for teams to implement them are devised by a few individuals.

The impact of good leadership and teamwork is an effective workplace with clear goals, direction and chain of command. Working in teams creates a social attitude towards success where both the group and individuals meet their goals. In addition, teamwork provides opportunities for feedback and elaborate communication – both facets of a pleasant work environment.

Important knowledge

Leadership

Having a good leader/boss is something all employees hope for. There are various ways to characterize this: A good leader is a person who has vision, is organized and enthusiastic, has a strong professional will and is able to take risks, although not too many or too large risks. A good leader can assess the future, has good vertical and horizontal communication skills and is willing to listen, respect, mentor and supervise the individual employees. It is also important that the leader is able to manage, inspire, motivate and create results. Familiarity and comprehension of technical aspects is necessary. In addition, a leader must have the courage to hire, fire and assign the right people to a particular task. Finally a good leader displays personal humility and knows and cares about his/her own traits and how they impact others. Leadership can be described as a function of all these capabilities. These capabilities constitute a good foundation, but do not guarantee optimal leadership.

Teamwork

Most people must work with others. Good teamwork is essential to a company's success. The parameters for good teamwork are many and resemble those of good leadership. In functional teams, team members show respect for authority figures by willingly following instructions given to them and providing deliverables on time.

> Team members must avoid challenging team leaders' authority

Guidelines for team leaders

1. Understand the strengths and weaknesses of the individual team members
2. Meet with team members regularly and get to know and *honor* individual team member goals
3. Understand how people and personalities may match or mismatch in the team
4. Obtain training in listening skills and conflict resolution and, if possible, allow team members to get similar training
5. When conflict occurs, find ways to resolve it quickly and assertively
6. Have frequent team meetings to ensure a high level of effective communication
7. Have a positive outlook and share it generously

Guidelines for team members

1. Understand own work style and capabilities
2. Trust others' capabilities
3. Understand that working with others is an opportunity to learn, teach and share
4. Be willing to teach each other new techniques
5. Share data and ideas with each other
6. Help each other troubleshoot and identify root causes of problems
7. When conflict occurs, find ways to resolve it quickly and assertively
8. Do not be afraid to share the credit for work with others
9. Have a positive outlook and share it generously
10. Enjoy working with other people
11. Enjoy working towards a common goal
12. Take responsibility for the team environment

Managing projects and people

Planning and executing projects is of the essence. This includes knowing what materials, tasks and deliverables are needed and at which time. Teams may perform pilot experiments and, after a successful trial, teach members to take over the function with supervision and oversight. When deviations from the plan occur, they must be reported and revisions must be made quickly. Often this involves time-consuming troubleshooting in order to identify the root cause of the deviation. Team leaders must be able to motivate and help team members through times of troubleshooting.

Being a good mentor requires being able to show interest in the mentored employee in a way that recognizes the needs of the employee. Being a good example to the employee and guiding the mentored employee to a deeper understanding is part of this work. Supervisors who take time teaching, listening and communicating can create a positive learning environment.

Colleague and peer interactions

A team is often a collection of people with varying amounts of education, experience and responsibilities. Team dynamics, therefore, spans many levels and requires good communication skills both vertically and horizontally. The interaction between colleagues and peers is different from that within such teams because here the communication is horizontal only. Some guidelines to becoming a good colleague and peer include the following action items:

1. Show interest and respect for colleagues and their project(s)
2. Interact with colleagues
3. Ask for help if it is needed
4. Offer help if requested
5. Avoid possessive or jealous behavior
6. Synergize.

Identifying with the company culture

No matter which level of responsibility an employee has (CEO, boss, mentor, peer, etc.) he or she needs to align with the company goals and identity. Working occupies a majority of a person's waking hours and being able to identify with what that work accomplishes helps lead to satisfaction.

Satisfied employees are likely to want to do well and perform their very best. When all identify with the company culture and goals, fewer obstacles to reaching those goals are created.

> Identification with the company goals requires knowledge and perception of working towards something larger than oneself

The first step towards identifying with the company goals is to obtain knowledge about them and one's deliverables. The second step is to commit to reaching the goals and producing the deliverables. So while reaching one's own deliverables, part of the company deliverables are met simultaneously. Perceiving oneself as being part of the company success rather than an individual success is a measure of how well one identifies with the company.

Anecdote: Leadership

Phil Moore had been promoted to CEO. This was his life's dream come true, to be in charge of a company, be its spearhead and its face of fame. Phil had replaced Will Modesty and took over a company in growth.

Will had led the company in the past ten years to incredible success. Will had achieved this by a belief that the best work should always be done, the highest ethical standards maintained and that certain risks should be taken in order to achieve higher standards. He was a humble leader with a professional will of steel and a heart of gold. The press had been very anxious to get interviews with Will and make him a superstar. Will declined this role and always gave others the credit for the good work: "The talented employees are the backbone of our success," he would say. Will had recently been headhunted to run a large international volunteer operation and he had accepted the challenge.

Phil thought Will was over the top in regards to avoiding personal publicity. His opinion was that if people perceived the CEO to be a superstar it would look good for the rest of the company. Phil invited the press; he called the marketing people and asked them to spend more time on advertisement, interviews, etc.

Phil's self-promoting attitude started to descend down the lines of the company. Directors who looked to Phil as a role model started to try to impress each other, team leaders would try to out-compete each other in order to impress the directors. Soon turf battles occurred concerning publication and patent authorships, ideas and innovations. Teamwork was not as pleasant any more. In addition, the employees started to be less innovative and take less initiative because Phil would usually take the credit for progress.

The situation started to deteriorate. Many mistakes were made because people were protective and possessive of their findings and communication levels and quality decreased dramatically, as did productivity.

Phil had to explain this to the Board. He had been given a successful enterprise and after six months it seemed like a disaster was imminent. He had no idea what had happened. A consultant who was hired by the Board to identify the problem concluded that Phil's focus on himself was a root cause of the crisis.

The Board decided to start looking for a new CEO. It was a very difficult search. Most people with the needed credentials were leaders just like Phil. Finally, the Board decided to coach Phil in leadership skills and give him a second chance to focus on the company instead of on himself.

Anecdote: Teamwork

Bing and Cherry were new team members. Both of them had recently graduated from university and for each of them this was their first job. They were not used to teamwork at all.

Ester Bond knew this would be an initial challenge when she hired them simultaneously. She was the team leader and would have to teach the new members to adapt as fast as possible.

The education came promptly. Management wanted a technical explanation of Ester's project, an explanation given by the workers in the laboratory. Both Bing and Cherry had been hired for the project in question. Each of them was responsible for different parts of the project. Ester asked them to work together on the management presentation.

The day for the presentation arrived. Bing and Cherry had had a lot of trouble agreeing on the content, what to present and which Powerpoint background to use. Each of them thought they knew best and that their project description would become less valuable if somebody else's ideas were added.

Ester decided to teach them a lesson. "Thank you very much," she said, when they handed over two different slideshows, one from each of them. "But unfortunately there is a problem because only half of what I need is in each of these slideshows. Bing, I want you to present Cherry's slides, and Cherry, I want you to present Bing's slides at the meeting," she said.

During the meeting it became clear to Bing and Cherry that because they both were responsible for the presentation, they both had to do well. Bing helped Cherry and Cherry helped Bing when it was necessary. Together they learned about the entire project and most of all, they learned to rely on each other as collaborators rather than competitors.

Summary of key skills for success

To be a successful leader you must be able to:

1. Lead
2. Serve
3. Deliver results consistently
4. Establish and contribute to company goals
5. Organize the company appropriately
6. Motivate people to do their best
7. Show and embrace optimism and humility
8. Harbor an enduring professional will for excellence
9. Align with company goals

To be a good team member you must be able to:

1. Be interested in others and their work
2. Understand your own personality and others' perception of you
3. Trust others
4. Share with others
5. Help others
6. Collaborate
7. Solve conflicts by negotiation
8. Be positive
9. Align with company goals

Mindset assessment: Leadership and teamwork

Leadership

Are you good leadership material? The following exercise will help you understand your current leadership potential. If your answer is yes to all questions, you most likely have the capability of becoming a great leader.

Question	A	B
Are you knowledgeable and *skillful* in your field of work?	Yes	No
Do you make consistent and productive contributions?	Yes	No
Do you contribute to group goals and deliverables?	Yes	No
Do you work well in group settings?	Yes	No
Are you able to organize people, resources and projects?	Yes	No
Are you able to meet set deliverables?	Yes	No
Are you able to stimulate commitment and excellence?	Yes	No
Do you show perseverance in attaining goals and visions?	Yes	No
Are you a person of humility and respect?	Yes	No
Are you a person of professional strength and perseverance?	Yes	No

Teamwork

Are you good at working in teams? The following questions and answers will help you identify your mindset in these two areas. If the majority of your answers are identical to Mindset 1, you can improve. If your answers are similar to Mindset 2, you are most likely a great team player.

Question	Mindset 1	Mindset 2
Would you rather work alone or with others?	Alone	With others
Have you learned about listening skills?	No	Yes
Do you employ active listening in your interaction with others?	No	Yes

(Continued)

Question	Mindset 1	Mindset 2
Do you trust other people?	No or sometimes	Yes, they can do as good a job as me if I supervise/educate them properly
Who should get credit for your work?	I should	Me, my team, my supervisor and the company
Do you socialize with your team members?	No or rarely	Yes, frequently or now and then
Are you comfortable assigning other team members to present your work?	No	Yes, after they have consulted with me so I make sure they have the latest details and the correct data
Do you discuss your ideas with other team members?	No	Yes, most often in order to avoid wasting my time and company resources
Do you identify with the company culture and goals?	No or I have not thought much about it	Yes, I enjoy being part of the company and work towards the common goals
How do you resolve conflict?	I try to ignore it	I try to resolve it as quickly as possible
Do you prefer to compete or collaborate?	Compete	Collaborate
Do you think others compete or collaborate with you?	Compete	Collaborate

Chapter

6

COMMUNICATION SKILLS

SCENARIO AND IMPACT

Imagine a world where communication suddenly is not possible. Most of all current activities would come to a halt. Communication matters in all environments in which messages are sent, received and interpreted. Companies are dependent on communications concerning company matters, work and work relations, and communication between all stakeholders.

The right number of messages must be created and **messages must be understandable**. Their reception must be assessed to determine whether or not **they are perceived the way they are intended**.

> Good communication = message is perceived as intended

It often requires additional correspondence to obtain clarity regarding both the intent of a message and the perception of the message after it has been sent.

> Communication can occur at different levels:
> in writing, in speech and in body language

Communication is performed at many different levels and as one-way, two-way or multi-way communication. One-way communication is defined as utilizing a medium in which it is impossible for other parties to create an immediate impact on the message. In two-way and multi-way communication, dialogue is possible.

> One-way communication is simple but its perception unknown

Examples of one-way written communications are reports, publications, instructions, applications, letters and faxes. E-mails and instant messaging can be classified as two-way or multi-way communication depending on the concurrent activity and the number of involved parties. Oral presentations, instructions and voice mail can be classified as one-way or two-way communication. Phone calls, discussions and personal meetings involve two-way or multi-way oral communication.

> Two-way and multi-way messaging is complex and provides knowledge about perception through feedback

Several personal factors influence both written and oral communication. These include: Communication style, level of listening that takes place, perception and level of conflict resolution. Finally, communication by body language can affect the message and requires that a "body" is present to deliver or receive the message.

IMPORTANT KNOWLEDGE

What are the rules of good communication? The various steps of the communication process are defined below.

Good communication rules

1. Send a sufficient amount of messages
2. Make sure each message is objective and clear
3. Improve knowledge about perception and understanding by requesting feedback

4. Use a positive and pro-active communication style
5. Employ active listening as part of communication
6. Use positive body language (if applicable)
7. Decrease ambiguity through additional correspondence
8. Display patience with the audience
9. See conflict as a creative opportunity
10. Decrease the potential negative impact of conflict through conflict resolution

Oral presentations

Mental state

For inexperienced presenters, it is common to feel anxious before an oral presentation. Other causes for anxiety include insufficient preparation, low self-esteem, fear of making mistakes or forgetting material and an expectation of humiliation. In addition, the audience plays a large role by their level of interaction. A very active audience can seem frightening to some presenters while others may perceive it as a sign of a successful presentation. Even experienced presenters may show signs of nervousness, especially if they are haunted by previous experiences.

To strengthen oneself mentally before giving a talk, the advice given in Chapter 3 on interview preparation can be used. Practice is essential.

Confidence comes with experience

Organization

Good presentations contain three main sections:

1. *An introductory description* of goals, main content and the relevance of the topic. The purpose is to secure the attention of the audience. A highly relevant hypothesis may be posed and the main points of analysis outlined.
2. *The actual "body" of the talk*, in which the content is explained and data/evidence is presented. This section is used for data description

and analysis. The presenter demonstrates his/her skills, credibility and capability.

3. *The conclusion*, in which the findings are summarized and prospects are presented. Finally, any contributing persons should be acknowledged.

Effective presentation materials

Slideshows are often the preferred presentation medium. Some basic rules apply to making a good slideshow:

1. Select a simple template and design slides well
2. Use slides to give main points only, use sparse amount of text
3. Create good transitions ("glue") between slides
4. Use a large font size (e.g., 24-point or larger)
5. Limit number of fonts
6. Highlight important messages or words
7. Use high quality and legible graphics
8. Respect copyrights
9. Remember references and acknowledgments.

Delivery

Before the presentation the speaker should make sure that the necessary equipment is functioning and compatible. It is wise to always bring back-up materials (hand-outs, flash drives, CDs, own computer, etc.) in case there is an unanticipated or unsolvable problem.

> Be prepared to present without the expected equipment

The first five minutes of a presentation are essential to success. This is the time to catch the audience attention, to show excitement about being on stage and to prepare the audience for a great experience.

> Start by thanking the organizers and the audience

A non-monotonous, enthusiastic and confident tone of voice tells the audience that the topic is interesting, of importance and worth hearing. A little humor goes a long way, although sincerity and credibility are also essential.

Moving around on stage should be done with purpose, to avoid creating a distraction or blocking the screen. Two of the most common mistakes are looking too often at the slides (on the screen) and reading directly from the slides. Doing so creates an impression of being unprepared or of having too much text on the slides. Most of all, it generates boredom among the audience. Instead of looking at the slides, the presenter should look at the audience. Once the talk has ended, the audience should be thanked for their attention and questions should be invited.

Managing questions and answers

One of the most feared parts of an oral presentation is the question and answer section. Many presenters worry that they will not be able to answer all questions and a lack of knowledge will be revealed in a humiliating fashion.

> No one knows everything;
> pretending to know is usually obvious
> and a lot worse than not knowing

Some pointers for good practice in a post-talk presentation are given below.

Guidelines for the question and answer section

1. Anticipate questions and prepare responses before the presentation
2. Listen carefully to questions and do not interrupt
3. Thank audience members for their questions
4. Reflect the questions back to ensure the question is understood
5. Do not panic if an answer is not obvious
6. Be honest
7. Do not pretend to know something, instead admit "I don't know" or maybe "That is a good question." This shows personal strength
8. If an answer is available, say it in a concise, clear manner
9. Move on to a new question when the time is right
10. End by thanking the audience for the questions and welcome additional discussions after the talk

Written presentations

Organization

Content, language and organization vary depending on the intended audience. The audience may be the general public (press releases, annual reports, etc., see Chapter 7 on Marketing), a specific scientific community (journal articles and other scientific publications), legal offices (intellectual property documents, see Chapter 12 on Intellectual property) or company officers (business plan, see Chapter 4 on Business goals and bottom line). As an example, the organization, effectiveness and publishing procedure for a journal article are described below.

In a **scientific journal article** the following components are usually required:

1. *Title:* The title should be formal, clear, short and descriptive.
2. *Authorship:* Authors are typically ranked according to contribution. Authors may be people involved in concept development, experimentation, analysis of data and writing of the manuscript. Who is included is often decided by the responsible author, whose name is often positioned first or last among the list of authors. Being the first listed author often means that this person did the majority of the work. As a guideline, authors should all be in understanding of, agreement with and responsible for the entire publication. All authors should have contributed academically to the publication.
3. *Abstract:* The abstract should describe the issue explored, the goals set and the main findings. One should be able to learn what the article is about by reading only the abstract. Typically, no references are listed.
4. *Introduction:* The introduction should describe the background information in the field and the void that necessitated the research or method which is described in the paper. Finally, the introduction should contain an outline of the main focus of the rest of the paper. The introduction is different from the abstract in that it provides a description of the background for the topic explored in the paper and includes key references.
5. *Materials and Methods:* In this section the actual prerequisites for the research are listed (materials) and the protocols used are presented (methods). No results are described here.
6. *Results:* Results are mentioned in this section. No analyses of results are given here unless it is clearly stated as an "analysis of results" or "results and discussion" section.

7. *Discussion:* This section describes the analyses of the results and the implications of these results. New prospects may become apparent through these results and give rise to new questions. Finally, it is common to include suggestions for future research based on the results.
8. *References:* The final section is a list of references. Journals differ in their format requirements and examples can usually be found in the publisher's instructions for authors.

Effective printed materials

Writing effectively is essential for the "selling" of a manuscript. Good writing uses well-organized information, proper grammar and correct spelling. To organize the content well means to be able to divide the available information into categories without repetition.

> Justify the presence of each sentence

Also, being able to tell an interesting story requires the ability to make good transitions between different information rather than a "dry" list of results. This can help to get the publisher's attention. Implications and extrapolations (not exaggerations) beyond the story itself make it even more attractive.

Editing and publishing

Selecting an appropriate publisher is sometimes a challenge. There may be many choices or very few. In cases of uncertainty it is a good idea to contact the editor-in-chief and submit the abstract to get feedback concerning the level of interest in the manuscript. Once the target journal has been selected, the manuscript must be formatted according to the publisher's guidelines.

Before submitting a manuscript it is advisable to let colleagues read it and suggest improvements. This is a good opportunity to correct writing mistakes. Be aware that it is difficult to please all people (but give it a try).

> Try not to take corrections personally;
> they are meant to improve the manuscript

After submission, the responsible editor will send the manuscript to a group of reviewers. The editor will compile their comments in a letter. It can take a while before the first correspondence appears. The manuscript may be rejected, accepted with major revisions, accepted with minor revisions or accepted. Changes are usually required. The author should carefully go through each of the reviewer's comments, comply with these or provide good reasons why the recommendations are not being followed and submit a revised manuscript as soon as possible.

When the manuscript is accepted by the publisher and it has been formatted it will be mailed to the author for approval. Once it has been approved and sent back to the publisher, the manuscript is sent to print.

Interpersonal relations

Active listening

Many conflicts arise because listening is seen as less necessary than speaking. Most people are not aware of the importance of listening. Listening can be hard work and there is a discipline called active listening. Good communication requires conscious effort in the area of active listening. Active listening means that the primary goals of the listener are to hear and understand the message that is sent by a messenger. Active listening means that the listener is *not* providing any judgment or advice, unless it is requested by the messenger. For many people, this is an unusual concept because they are used to evaluating and judging the content of the message immediately as well as giving their opinions and advice.

> The listener's **only** task is to fully understand the messenger

The following steps can be used to describe active listening:

1. *Encouragement:* The listener encourages and asks for information in a gentle manner: "I am very interested in learning about what you think of/what happened . . . will you tell me what you think?"
2. *Reflection of content:* The listener reflects on the received information and gives the messenger an opportunity to correct the perception. "What you are saying is . . . and I hear that it means . . . Did I understand what you said correctly?"

3. *Reflection of feelings/conclusions:* The listener attempts to relate to the message, interpret the impact of the information and give the messenger an opportunity to correct the perception: "It sounds like you think/feel that . . . That must be great/awesome/hard/bad . . . is that what you meant and how you feel/what you want to do?"
4. *Summary:* Further clarification is attempted and a conclusion is proposed. The listener tries to summarize the situation and gives the messenger an opportunity to correct the perception: "So, is it correct that you think that . . . and because of that you feel . . . and in light of that you would like . . . ?"

> Active listening requires 100% attention towards listening; it is neither a discussion nor a place to voice the listener's opinion

Taking the time to actively listen means that there is an opportunity to create a "safe space" for the messenger to voice a concern and share an event or an opinion. Active listening is an opportunity to minimize second-guessing and ambiguity. Active listening is a means to display a sincere interest in understanding, to make others feel listened to and understood, and to create a mature basis of respectful communication.

When active listening becomes a routine and an integral part of communication and later discussions, the communicating parties will alternate performing roles of messenger and listener.

> Conversations that employ active listening are more likely to conclude in synergy rather than antagonism

Active listening is best performed as an oral interaction face-to-face, but the basic principles of displaying a will to understand the message can easily be applied to written communication.

Communication styles

A **poor communication style** may create compliance, but the way it happens is highly unpleasant. Examples of poor communication styles include judgmental statements, manipulation, orders, threats, preaching, interference, blaming, accusing, personalizing, assuming, prying, use of words with negative connotations and poor language.

A **good communication style** can be defined as communication that creates both commitment and compliance. It creates a safe and positive environment in which active listening can take place. Good communication may include praise, encouragement, use of words with positive connotations, assertiveness (see below), interest in listening to others and interest in their differences.

An **assertive communication style** is characterized by the conscious use of "I-statements" and the avoidance of "you-language." Being assertive means that a person expresses his or her own wishes, takes responsibility for his or her own feelings and that the person suggests solutions to his or her own complaints. A standard display of assertive communication looks like this:

"I-statements"

1. When . . . *my project was terminated*
   (describe **what happened**)
2. I felt . . . *upset and hurt*
   (describe **what you felt**)
3. I felt this way because . . . *I enjoyed my project and I did not understand that it was not making good business sense to the company any longer; I was taking it personally*
   (describe **why you felt that way**)
4. In light of this I would like . . . *to be more flexible and not take things so personally when projects have to change*
   (describe **your wishes** based on the above statements)

Conflicts

Conflicts range from creative opportunities such as opposition of ideas and/or principles and simple disagreements, to damaging interactions such as controversies and, at worst, fights. Some conflicts, therefore, are positive and lead to constructive thinking while others are entirely destructive.

The **symptoms** of destructive conflicts are many and easy to recognize: Lack of effort at active listening, impatience, lack of assertion of wishes and possible solutions, "you-language," provoking words, inappropriate advice, possessive behavior, inappropriate personalization of problems, etc.

Table 6.1 Conflicts and conflict characteristics

Conflict impact	Conflict type
Creates innovative thinking	Constructive
Optimizes decision-making	Constructive
Accepts disagreement	Constructive
Accepts challenge	Constructive
Is viewed as important and natural	Constructive
Damages interactions	Destructive
Enhances possessive behavior	Destructive
Happens in a coveted manner	Destructive
Creates low self-esteem	Destructive
Decreases level of trust	Destructive

Conflicts may be **caused** by many factors, including different value systems, different personalities and, in general, by diversity. Power struggles are often the common cause of conflict.

The **style of conflict resolution** varies depending on the personality and interpersonal skills of the involved parties. If the style is highly competitive, there is a strong wish for an individual "win" situation and this can hinder a "win–win" scenario for both parties. The conflict may not be solved, rather a power struggle will take place. In contrast, if the style is accommodating or avoiding, the parties involved in conflict send the message that all conflicts are hurtful and should be avoided. The opportunity to be creative and learn about potential improvements and innovation may be lost. In addition, conflicts are not truly resolved but rather postponed and accumulated. The optimal conflict style is collaborative and compromising. In this type of conflict, solutions are created while a win–win situation is being sought.

> Collaboration and compromise are essential for creating win–win scenarios

Conflict resolution requires great skill and consists of several steps:

1. *Identification of symptoms and parties involved in conflict:* The symptoms of conflict that have created the conflict must be identified. The involved parties should be brought together for a meeting.
2. *Description of situation:* The parties alternate roles of messenger and listener. The messenger needs to attempt to use assertive communication and the listener to display active listening skills.

3. *Identification of causes for conflict:* Based on the communication between the parties, the cause(s) of conflict should be determined.
4. *Plan for resolution:* Together the parties should establish a collaborative and compromising resolution style and work towards a plan that embraces a "win–win" situation.
5. *Agreement and commitment:* The parties should be in full agreement with the plan and be committed to its execution.
6. *Level of satisfaction:* Satisfaction with the outcome of conflict resolution must be confirmed by all involved parties before the issue can be considered settled.

Anecdote: Interpersonal relations

Max had just come back from a two-week training session about communication skills. Nobody recognized the way he now was interacting with his employees, but this was a good thing.

Previously, Max had run into problems with his group. They had shown up at Bridget's door (she was the HR Director) and told her that they would all quit because they did not want to work for Max any more.

She had called a meeting with Max and his group. Max's group consisted of highly skilled experts in their fields. A few other people were invited as observers and listeners. Max's group manager, Larry, had been asked to describe what had happened.

Larry told the audience how hard the group worked for Max. He also described that no matter how hard they worked, their perception was that it seemed never to be good enough because Max tended to use very demeaning and negative language. Larry described how this affected the group and his own work. He specifically talked about a time when Lauren had made a mistake. When Max learned about the mistake he came into the lab and started blaming her and the group for doing a poor and unacceptable job: "All I get from you is incompetence." Max also used demeaning language and talked to them as if they were school children, not adults: "You need to go home and do your homework," he said. Furthermore, he displayed an unrealistic level of expectation by adding "Making mistakes is just not acceptable!"

Lauren immediately went to Larry with her resignation. Larry said that the group had anticipated Lauren's resignation. Max had gone too far; he had called them incompetent. He acted as if he was considering mistakes totally unacceptable. Larry finished his description.

Max had had a hard time during the meeting. He was constantly interrupting Larry, saying, "This is not true, they just do not do their jobs well."

Bridget sent Max to "communication school." When he came back Larry and his group (including Lauren) were called to a new meeting to once again discuss what had happened. This time, Max sat quietly and listened to Larry and the group. After they had spoken, Max reflected on the event. "What I heard you say was (1) a mistake was made in the lab; (2) you did everything you could to fix it; (3) I came into the lab and behaved inappropriately in a way that felt demeaning and threatening to you; (4) Lauren quit and you decided to do the same. Is this a correct reflection of what you described?" Larry nodded. Max continued to reflect on how the event had made the group feel. "You felt that I was impossible to please, and not at all nice to work for. Did I understand this correctly?" Again Larry nodded.

This was a breakthrough. Max had been able to listen to the group and their feelings without attempting to judge the content, deny it or give advice. In addition, Max took time to apologize to the group. "I never understood how I communicated or the impact my communication style had before I had the training," he said. "Once I understood this, I felt terrible about how I have been speaking to you, and I really want to try to do better. I am very sorry about what has happened in the past. I hope that you will accept my apology and give me a second chance."

Days went by and things became much better. Even Lauren came back and Max worked respectfully with her to earn her trust again.

Summary of key skills for success

To make a successful oral presentation you must be able to:

1. Prepare good presentation materials
2. Prepare yourself until you feel confident about the content
3. Prepare yourself mentally
4. Prepare questions and answers
5. Be excited about interaction from the audience

To make a successful written presentation (journal article) you must be able to:

1. Write well
2. Organize your data and writing
3. Identify an appropriate publisher
4. Follow publisher's guidelines
5. Take criticism professionally

To successfully relate to other people you must be able to:

1. Actively listen
2. Send clear messages
3. Use pro-active, assertive "I-statements"
4. Check to see if your messages are perceived and understood correctly
5. Use positive body language
6. Be patient
7. Identify the symptoms of conflict
8. Distinguish between constructive and destructive conflicts
9. Be pro-active and interested in solving conflict

Mindset assessment: Communication skills

How well do you communicate and relate to others? You can get an idea of areas in which you can improve by relating to the following two mindsets. Mindset 1 is harbored by a person with good communication and interpersonal skills, whereas Mindset 2 is found among people with poor skills in these areas.

Mindset 1	**Mindset 2**
An oral presentation is an exciting opportunity to tell a story and get the audience interested in what I am doing	Oral presentations are challenging and I often feel judged and interrogated
I am grateful for questions from the audience because the audience lends intellect and thoughts to me which I did not have	I dislike the question and answer section. It is often embarrassing, especially if I cannot answer the questions
I like to write about my work and present it in a palatable manner to a professional publisher	Writing is necessary so I can prove what I did
The reviewers usually have great suggestions for improvements	The reviewer's comments are often unreasonable and create a lot of work for me

(*Continued*)

Mindset 1	**Mindset 2**
I generally use words which are objective and have no emotional connotation	I think it is important to emphasize my feelings by using words with strong emotional connotations
I prefer the use of "I-statements"	I like to point out when somebody has done something wrong by using "you-language"
I quickly identify conflicts	I am not really aware of a conflict before I am in the midst of it
I will control my emotions during a conflict	During a conflict, I tend to get upset or angry with people
During a conflict I try to perform active listening in order to fully understand the conflict and what started the conflict	I think it is important for everybody everybody to hear my opinion during a conflict in which I am involved

Chapter

7

MARKETING

Good marketing is achieved by creating a favorable impression of a "product" – an item, an idea, an environment, a status or situation, anything that can be considered desirable and obtainable. Success in marketing endeavors depends on the ability to make others interested in the "product," and in spending time convincing them to discuss, negotiate and finally "buy" the product. Everywhere "products" (what people can do and offer) and people's impact on the "products" (skills, personality) are marketed. The better the marketing effort is, the larger the likelihood for success. In contrast, if no marketing effort is made, products will remain largely unknown.

SCENARIO AND IMPACT

There are two main avenues for marketing: The company visage towards the outside world and internally among individual employees.

> Marketing efforts are needed internally as well as externally

All companies need successful marketing of product portfolios. New companies, which have yet to establish products, need marketing to create an attractive environment around their ideas and prospects for future products.

The audiences are many: The general public, investors, employees, other companies, etc.

There are several ways to market a product. This chapter will focus on both written marketing vehicles (press releases, annual reports and web pages) and oral performances (meetings and networking events).

Start with the outcome in mind

There are several key elements to consider when performing effective marketing: The actual **marketing vehicle, audience, confidentiality, content, communication style** and **tone**, strategic **wording, length** and **anticipated outcome**. Awareness about these elements can help identify the best signals to send to an audience.

Important knowledge

Signal and response

Sending signals that solicit the predicted outcome is part of a successful marketing effort. The desired response spans from attraction and curiosity to sincere interest and "purchase." In order to attract people and incite curiosity, a message creating **excitement** and a **competitive advantage** must be given. The message must have **substance, importance, implications** and be **trustworthy** to create sincere interest.

A marketing message must catch attention, be truthful and have important implications

In the following example, a researcher has discovered a new medicine and the company wants to send out a press release. The researcher is asked to help the marketing group to write it. Below is a set of notes/ideas and guidelines from the marketing group to the researcher:

Excitement: A new and/or unusual discovery and/or the potential extinction of a serious disease (please name it). Write in an optimistic and opportunistic tone.

Competitive Advantage: Why is this medicine better? (There was none before for the particular disease.)

Substance: The concise laboratory findings are described here followed by a brief description of how the findings were conceived. This part should be short and to the point. Confidentiality must be observed.

Importance: Write about how important the disease is, how many have it (?), the medicine will be saving people, money and environment, avoid spread of disease, few or no side effects.

Implications: Saving people from a serious disease, no former cure, now creating a cure, making it possible to get prophylactic treatment as well, low cost.

Trustworthiness: Only truthful statements which can be defended if necessary should be given. No compromises. Wording may be strategic and contain several less concise descriptors such as "may cause," "in progress," "part of," "some effects," etc., where needed (if any uncertainty).

Maintaining a tone of optimism and opportunity will help the reader to believe that there is a wealth of prospects. Words with positive connotations rather than neutral or negative connotations should be utilized to further secure this outcome. Likewise, negative statements should be converted to positive ones.

Use positive statements

Observations should be described in an objective manner to maintain trustworthiness. Facts should be stated in a non-generalizing (unless it is a general fact) and non-exaggerated manner.

The guidelines above apply to many kinds of marketing and communication efforts, including the effort used to present people to each other. It is relevant whether it is conducted through a written or oral communication medium. Knowing how to use the different marketing vehicles appropriately is essential to both company and employee success. Examples of some of the most common parameters for each of the vehicles are mentioned below, but these do not exclude additional features.

Press releases

Writing a press release correctly is of utmost importance for a company. A press release is like a momentary glimpse of part of the company life.

It contains a brief description of the latest progress, usually a single event. The event could be a merger, a contract, a patent, or a new collaboration agreement, to mention just some examples. It is a public statement that most likely appears on the Internet (company website) and can also appear in various status reports. The press release archive is useful for surveillance of company dynamics.

A press release usually provides insight into a single event

The purpose of a press release is to inform the public about company news and to positively influence public opinion. Press releases are carefully monitored by competitors and stock brokers (if the company shares are publicly traded).

Press release parameters

1. *Audience:* General public, stakeholders, competitors, stock brokers
2. *Confidentiality:* Observed
3. *Content:* News and events: financing, mergers, acquisitions, leadership changes, patents granted, new collaborations, contracts, publications, general information about the company
4. *Communication style and tone:* Brief, seemingly objective yet positive, milestone oriented, opportunistic, optimistic
5. *Wording:* Legally correct, strategic such as "has the potential," "has promise," "may provide," etc.
6. *Length:* Short, often less than one page
7. *Anticipated outcome:* Positive influence on stock brokers, market, showing dynamic company with great progress

Annual reports

Annual reports show an overall picture of a company's culture and life. The reports contain information about the company's mission and business goals, timelines, organizational structure and the progress made throughout an entire year. Sometimes prospects and budgets are included as well.

The report is written to quickly give the reader a favorable impression. It is written once per year.

> Annual reports provide a quick overview of the entire enterprise and events in a calendar year

Writing these reports in a manner that integrates all positive contributions and assets is a large task. Many people are usually involved. The report contains easy-to-read information and may include pictures; it is therefore a handy advertisement tool.

Annual report parameters

1. *Audience:* Anybody with interest in the company, employees, potential partners, collaborators, job seekers
2. *Confidentiality:* Observed
3. *Content:* Description of the company, its history, mission and business goals, organization, areas of work, publications, patents, collaborations and anything else which is considered non-confidential and of market value
4. *Communication style and tone:* Descriptive but not detailed, many pictures, "pretty," positive to neutral presentation
5. *Wording:* Legally correct, concise language
6. *Length:* Booklet, length variable
7. *Anticipated outcome:* Makes company look organized, financially well-managed, list of progress is intended to impress and pique curiosity and interest

Web pages

Company information on the Internet can be found on the company's website. This site usually contains a wealth of general information. In some cases, the information may resemble the content of the annual report. One clear difference is that websites are often updated more frequently than annual reports, and, therefore, the information is more recent.

A company website contains updated public information about many aspects of the company

A website may also contain hyperlinks to a press release archive, even prior annual reports, publications, patents, job openings and other relevant information such as stock quotes (if the company is publicly traded) and contact information. This medium is likely to contain the greatest amount of public information available. The purpose of the website is to create interest, post updated information for all stakeholders and often to provide an on-line application opportunity for job seekers.

Website parameters

1. *Audience:* General public, all stakeholders
2. *Confidentiality:* Observed
3. *Content:* Similar to press releases and annual reports plus additional information (often hyperlinked) about jobs, patents, etc., contact information and directions to the company
4. *Communication style and tone:* Similar to press releases and annual reports, some details (maybe through hyperlinks)
5. *Wording:* Promising, neutral to positive
6. *Length:* Unlimited, depending on amount of information and hyperlinks
7. *Anticipated outcome:* Interest in browsing website, interest in company, on-line job applications, inquiries of various kinds

Meetings

A meeting between employees within a company is an internal meeting. If this meeting takes place between employees who do not know each other, it may be considered an internal networking event. Such events could include a company reception, a celebration, or a picnic. Internal events take place either in the company or in settings where only company employees are present.

Employees need to market themselves to each other; personal and professional reputation determines success

There are many purposes of internal meetings and networking events. They include discussions about (confidential or non-confidential) matters, marketing of new ideas, to the creation of new networks within a company, and interest in other employees' work.

Internal meeting and networking parameters

1. *Audience:* Employees participating in event
2. *Confidentiality:* Participants are employees and can usually share confidential information
3. *Content:* Any matter relating to company business meeting or general and introductory matters if this is a networking event
4. *Communication style and tone:* Polite, honest, inquiring, opportunistic, enthusiastic, interested, frank
5. *Wording:* Professional, not legalistic
6. *Length:* Any length of meeting (lunch: often one hour) or event (picnic or reception: a few hours)
7. *Anticipated outcome:* Marketing of new or existing effort, planning and creating support for the future of internal activity, interaction between employees from same or different groups, strengthening of company culture and opportunity for interdisciplinary thinking and invention

Meetings and networking events between employees and non-employees are external events. These events may be planned or may happen spontaneously. They may take place within the company and be based on invitations to non-employees or they may be physically located elsewhere, either in public or non-public venues. Public venues include conferences, receptions, workshops and grant panels. If the event takes place between two or more companies, the meeting is a non-public event.

The purpose of external meetings and networking events is sometimes to create new contacts and interest with potential collaborators, competitors, partners or other parties. Other times, the purpose is less clear and spontaneous interactions may be the primary driver for new networks.

External meeting and networking parameters

1. *Audience:* General public, networking event attendees, depending on venue
2. *Confidentiality:* Observed
3. *Content:* Depending on topic, but generally relating to event, non-confidential information about work areas
4. *Communication style and tone:* Enthusiastic, inquiring, interested, polite, honest, opportunistic
5. *Wording:* Legally correct
6. *Length:* Any, often one hour, length of individual conversation varies, length of events vary
7. *Anticipated outcome:* Creation of interest in interaction, collaboration, new contacts and ideas

Press conferences

Companies that are publicly traded have obligations to share their results regularly. Press conferences serve as an outlet for the display of the present company status. Commonly, the CEO and CFO will present the details. These conferences can sometimes be attended by selected employees and the public may be able to listen to live web casts.

> A press conference is a verbal display of current results tailored to instill consumer confidence

How well a press conference is received can have a large impact on the company's reputation, stock value and future. Holding press conferences at regular intervals ensures that the public stays informed about the most important issues in the company.

Press conference parameters

1. *Audience:* General public, all stakeholders
2. *Confidentiality:* Observed

3. *Content:* Introductory speech about matters relating to company status and progress, such as financial data, future strategy, etc., answers to questions from the general public, stock brokers, etc.
4. *Communication style and tone:* Polite, honest, opportunistic
5. *Wording:* Legally correct, professional
6. *Length:* Usually 30–60 minutes
7. *Anticipated outcome:* Verification of company's health and future prospects, opportunity to interact directly with stakeholders and address their eventual concerns

It is of universal importance that people conduct themselves with engagement and excitement when using any of these marketing vehicles. With the outcome in mind, a message of substance with a positive focus must be presented. A good marketing effort fosters belief in importance of the work and that the person conducting the marketing is trustworthy.

Anecdote: First impressions

Ann worked in the marketing department of a relatively new and highly successful, but largely unknown company. She was attending a networking event at the local industry center. She had been asked to present her company's goals and major strategy for the attendees. They were mostly industry executives from the area. She did not know many of them yet.

Ann dressed professionally as always and was careful not to wear too much make-up or perfume. Her business cards had been updated and she also brought the company's annual report with her. She had been practicing a small talk yet she was going to try to seem somewhat spontaneous and lively on the stand. She was very excited about the opportunity.

After the initial registration event a small reception took place. Ann enjoyed this part; she smiled, greeted people, learned their names and took time listening to their stories. She was good at making people feel comfortable in her presence; her demeanor was confident, calm but enthusiastic. She gave out many business cards that day.

After the reception Ann gave a splendid presentation and answered a lot of questions. She was very professional and did not get upset when somebody said that they had never heard about the company for which she worked. She was well prepared mentally and handed out the annual report with confidence.

The next day she received several phone calls:

A journalist wanted to interview the CEO of the company for which Ann worked; he had become excited and interested in the company by listening to Ann.

The VP of research from a company close by wanted to inquire about a potential collaboration. It seemed like the technical part of Ann's presentation was well understood and had plenty of substance and interest.

Also, several competitors wanted to know more about the company product line. Ann had made it very clear that the new products would have great implications for the current market.

Finally, somebody wanted to hire Ann because she had presented herself and her company so well. She had shown great enthusiasm and professionalism. This was quite a surprise for Ann and she felt very flattered.

It was clear that not only was Ann very talented, she was also able to use her talents in a way that served her company well. Ann was promoted quickly and stayed with the company for several years.

Summary of key skills for success

To use the different marketing vehicles well you need to:

1. Know what kind of outcome is intended
2. Know the audience
3. Know what is confidential
4. Know when confidentiality must be observed
5. Understand the content that is marketed
6. Use the best communication style
7. Select the appropriate marketing vehicle
8. Use professional and legally correct language
9. Create a message that is positive, captivating, truthful and important
10. Find the best avenues to disseminate the message
11. Update all messages regularly
12. Take feedback seriously and adjust messages accordingly
13. Develop good interpersonal skills
14. Network

Mindset assessment: Marketing skills

Do you know how to market yourself and the company for which you work? Below are some selected statements you can use to assess whether you find marketing difficult or easy. They are by no means meant to encompass all necessary marketing skills. If you tend to agree with most statements you are most likely a great marketing talent. If you tend to disagree, you may have a different style or persuasion technique or simply need to work on a few issues if you want to improve the outcome of your marketing efforts.

Marketing yourself (anywhere):

Statement	Tend to	Tend to
I have a positive and kind demeanor	Agree	Disagree
I smile and make eye contact when I speak to a person	Agree	Disagree
I am interested in the people to whom I speak	Agree	Disagree
I listen carefully to what people say	Agree	Disagree
I think about what I am going to say before I speak	Agree	Disagree
A conversation involves people both speaking and listening	Agree	Disagree
I am confident when I present a message	Agree	Disagree
I can interact with people at all levels	Agree	Disagree
I am enthusiastic when I network with others	Agree	Disagree
I like to meet new people and get to know them	Agree	Disagree
I do not get upset when people challenge my opinions or messages	Agree	Disagree
I stay calm when I am under pressure	Agree	Disagree
My messages create interest	Agree	Disagree
My messages are truthful	Agree	Disagree
I listen when people criticize what I say, then adjust my message	Agree	Disagree

Marketing the company you work for (professional settings):

Statement	Tend to	Tend to
I think about my audience before I create my message	Agree	Disagree
I think about the outcome first	Agree	Disagree
I like to present a well-prepared talk	Agree	Disagree
I do not mind presenting with short notice	Agree	Disagree
People tell me I am good at presenting my messages	Agree	Disagree
I honor confidentiality	Agree	Disagree
I know the balance between confidentiality and creating interest	Agree	Disagree
My messages have substance, importance and a competitive edge	Agree	Disagree
I update my messages regularly	Agree	Disagree
I know the difference between different marketing vehicles	Agree	Disagree
To my best knowledge, I use professional and correct legal language	Agree	Disagree
I am able to write a press release	Agree	Disagree
I am able to contribute to the annual report	Agree	Disagree
I provide information to our website as needed	Agree	Disagree
I take the opportunity to network when I can	Agree	Disagree
I know what other groups in the company are doing	Agree	Disagree
I know what other companies are doing	Agree	Disagree

Chapter

8

DISCIPLINE

Practicing discipline means following a set of guidelines, rules or regulations. Most environments require a certain level of discipline, which may be succinctly expressed or unspoken, and industry is no exception. The more optimal the guidelines, and the better the discipline of the employees, the more likely the company is to succeed.

Discipline is a **must**

In order to create a platform for success, employees must be made aware of the nature and level of discipline that is expected of them.

SCENARIO AND IMPACT

The need for discipline is found throughout an organization at both the individual and team levels and relates to all conduct necessary to achieve determined standards. There is a certain level of basic discipline needed to set goals for production (technical and managerial measures) and work environment (personal measures) in regard to producing deliverables on time, maintaining quality of deliverables, managing projects and people, documenting effort and conducting oneself appropriately (see below and also Chapters 12, 14 and 15).

> Discipline at three different levels is needed: technical, managerial and personal

The exact actions needed depend on individual roles and placement within a company. A company must have visible guidelines and expectations for discipline that create a culture of accountability. In this way, individual employees have incentives for holding themselves to a certain standard of both professional reliability and sociability.

IMPORTANT KNOWLEDGE

Deliverables

Employees must understand their exact deliverables, the necessary quality of these deliverables, and produce and deliver them in a timely fashion.

> Know how to produce the expected deliverables and deliver them on time

Some examples of deliverables are: a business plan, a market analysis, patents, inventions, production of vaccines or new contracts. The area of work depends on the employee's role in the company and must be understood along with the tools and protocols/procedures (if any) that are going to be used. Each step of the work, procedure or analysis must be performed with great care and consistency and all data must be recorded appropriately.

> Be careful and consistent

Deadlines and punctuality

A deadline is a time-point at which certain tasks or projects need to be finished and a product delivered. The deadline is often non-negotiable and must be respected.

Deadlines are often non-negotiable

It may seem like common sense to arrive on time for a meeting. A meeting time can be considered a deadline to show up. Certain people and cultures have an easier time complying with deadlines and are more used to punctuality than others. It is important to display that punctuality matters in order to establish respect for people's work and time. Being late can cause disruption and a decrease in effectiveness leading to a loss of resources for the company.

Another way of displaying respect (and discipline) for others' time is through preparedness. A chance to present one's work may not come along often and it is essential to give a good impression.

Show respect for others' time by being punctual and prepared

Project management

As a precursor to beginning any kind of work, the work needs careful thought and planning. The work may relate to any industry matter such as laboratory experiments, production of articles, analysis of a market potential, production of an annual report and many other areas. When companies grow, it is common to have specialized project managers who concentrate on obtaining on-time, quality deliverables from the various groups. There are several software tools for project management. Choosing which one to use may depend on what is already in use, accessible and interactive at the company.

Think, plan, then act

If projects do not go as planned this is most often because unexpected change happens. When change occurs, the planning needs adjustment. Planning and project management is therefore an integrated and ongoing process, initiated as a plan and continuing as surveillance with alteration/adjustment of the plan.

Record keeping

Good management includes collecting and recording data properly and on a regular basis. In research laboratories employees typically are instructed to use company notebooks and to record experimental plans, expected results, actual results, analysis of the results and further planning. One of the reasons this is so important is that these notebooks are part of the evidence for invention dates (see Chapter 12). Employees who work in manufacturing of foods and drugs must comply with a rigorous set of regulations to document their work (see Chapter 14).

> Errors must also be recorded

Good records contain all kinds of results and also errors that may be analyzed for future learning. Scientists who perform research know that only a fraction of their experiments lead to useful data. By being aware of errors and experiments that lead to unexpected or no results, future errors can be avoided. Unexpected results can also become a resource for future invention.

Personal conduct

Discipline at the personal level has a large impact on the work environment. It is often underestimated and not discussed much. When people conduct themselves properly, it has a positive impact. In contrast, it does not take long to destroy a good environment by introducing bad habits and misconduct. There are a wealth of issues to consider while describing proper conduct and personal discipline:

1. *Be truthful.* There are different ways to tell the truth. It is recommended to display assertiveness while remaining tactful.
2. *Report errors.* Many people are afraid to report mistakes or errors. Avoiding doing so creates poor communication. Errors can cost companies much money if they are not realized or reported.
3. *Take responsibility.* People make choices about the way they present themselves and interact with others. Understanding one's own habits helps in understanding interactions with others.
4. *Be accountable.* Apologizing quickly when somebody was hurt by words or if a mistake was made can help solve conflict in a timely

manner. The ability to admit mistakes can be contagious and is a good example for others.

5. *Be positive and helpful, avoid negativity.* A positive atmosphere promotes creativity and growth. There is virtue in staying positive, especially through challenging times. Because negativity has the potential to create a sense of dissatisfaction, fear and tiredness, it indirectly demands many resources and should be avoided.
6. *Show respect and appreciation for others.* Feedback promotes understanding of behavior. When positive outcome occurs there is an opportunity for positive reinforcement by appreciation.
7. *Listen to others.* Active listening (see Chapter 6) is a powerful tool to promote well-being and respect among employees. Active listening is a demanding skill, which requires staying objective, empathetic and non-judgmental during an interaction. It is a huge benefit for a company to have employees trained in this discipline.
8. *Work with consistency both individually and in teams.* Being eager to perform well alone and in teams is part of being a good citizen within a company. Team members rely on everybody's efforts and all team members should be aware of their expected individual contributions to the team.

Anecdote: Delays

Heather and Sam were selected to give presentations at the executive board meeting. Heather was a scientist in the lab. She always had new great ideas and her newest one inspired management to ask her for a 5-minute presentation. She knew the business development team, including their newest hire, Sam, was going to be present and that they were going to analyze her idea.

The day of the meeting arrived and the Executives started appearing in the board room. The CEO, the VP of research and the head of corporate counsel were among the participants. At 9 a.m. everybody was waiting for Sam. At 9.15 a.m. the CEO prompted everybody to start without Sam. Heather did a great job and harvested praise for her ideas and well-prepared presentation. Sam's colleague Bill was asked to stand in for Sam. Bill was not prepared for this. Finally, at 9.30 a.m. Sam showed up.

Sam was very lax about the meeting time and this was an unfortunate behavior. The business analysis was almost over. Sam, who had done a great amount of work of high quality, was mostly remembered as the person who made everybody wait. Bill was remembered as a stand-in who was ill prepared.

Sam's supervisor was asked to make an effort to change Sam's attitude towards meeting on time. The supervisor made Sam calculate the cost of having all the people wait on him by adding up the salaries of the CEO and all the participating Executives, Heather's salary, as well as the cost for letting the meeting exceed the projected meeting time. Understanding that time is a commodity and that the delay cost the company thousands of dollars in salaries and lost effort, Sam was never late again.

It took a while before both Sam's and Bill's work was taken seriously and evaluated objectively by colleagues who had attended the meeting.

If it is not possible to be on time for an event there needs to be a valid reason for the delay and a notification must be given before the delay occurs. Being aware of the possibility of a delay by having a prepared stand-in is a good strategy.

Summary of key skills for success

To display discipline you need to:

1. Know your deliverables
2. Obey instructions
3. Produce quality work consistently
4. Record data appropriately
5. Obey deadlines
6. Be punctual
7. Contribute to a pleasant work environment

Mindset assessment: Discipline

You may try to address the sample statements in the following three areas to understand if your level of discipline is sufficient for success in your current situation. If you find yourself in agreement, you are probably a highly disciplined individual. If you disagree with most statements, you may want to adjust your work style.

Discipline in a laboratory environment

1. I familiarize myself with equipment before I use it
2. I read and understand protocols before I start a procedure

3. I never deviate from a protocol and never take shortcuts. I am mindful of each step in the protocol
4. I take great care while performing technically
5. I always use a timer if instructions specify timing
6. I always include controls in my experiments
7. I record data immediately and I never ignore data
8. I notify my supervisor if I make an error
9. I notify my supervisor if something looks aberrant
10. I make sure my work is consistent in terms of quality and amount

Discipline in a business development environment

1. I familiarize myself with the market before I analyze it
2. I read about and understand our competition before I make any predictions
3. I never ignore any data
4. I am mindful of each step in my analysis and predictions
5. I take great care while performing these tasks
6. I always use appropriate databases and software
7. I record data immediately
8. I notify my supervisor if I make an error
9. I notify my supervisor if something looks aberrant
10. I make sure my work is consistent in terms of quality and quantity

Work discipline (general)

1. I familiarize myself appropriately with the area of work
2. I understand the area of work and the tools I need to use
3. I follow procedures
4. I am mindful of each step
5. I take great care
6. I record data immediately
7. I notify my supervisor if I make an error
8. I notify my supervisor if something looks aberrant
9. I make sure my work is consistent in terms of quality and quantity

Managerial discipline

1. I arrive before or on time for work and meetings
2. I make sure equipment works before I need it
3. I meet my deadlines and deliverables
4. I plan my work before I do it

5. I review and record data immediately
6. I analyze obtained data prior to subsequent planning
7. I regularly meet with my supervisor to discuss progress
8. I regularly meet with a colleague to read and sign notebooks
9. If I make an error I report it immediately

Personal and inter-relational discipline

1. I tell the truth
2. I report my errors
3. I feel responsible and accountable for my work
4. I am kind and helpful to others
5. I respect and acknowledge my colleagues
6. When in a team, I contribute my part to the team's work and business goals
7. I practice active listening skills
8. I work on improving my interpersonal skills

Chapter

9

FLEXIBILITY

Change is inevitable. Resistance to change is common and is often used to protect valuable habits, rules and projects. A dynamic interchange between the pressure for change and the resistance to change helps establish an appropriate middle ground between preservation and renewal.

> It is not a question of if, but when, why and how change happens

SCENARIO AND IMPACT

Change happens everywhere. In a dynamic environment change happens at a higher frequency than in a conservative environment. Industry is typically a highly dynamic environment. There are many different triggers of change in industry. One example of these is altered market forces, which can lead to both positive and negative changes. These changes are likely to cause revision of priorities and could induce company reorganizations.

It is easy to understand that if there is no change within a certain environment, then there is neither adjustment to nor interaction with a surrounding or adjacent dynamic environment. In addition, because there is no adjustment there can be no improvement.

> Change is necessary for improvement

Adaptation to new situations is highly pertinent and the speed with which the adaptation takes place is of the essence. Seeing change as more than a necessity, but an opportunity for improvement, will help accelerate the process and positive expectations, even when a change is initially viewed as negative.

> Quick adaptation to change increases likelihood for success

Large companies or companies who have maintained the same employees and projects for a long time may be less dynamic than their smaller counterparts. Companies that adapt fast may be able to respond in a more timely way to altered market forces by redistributing their resources and changing their priorities. The impacts of these dynamics on the individual and society are highly visible; employees may keep their jobs and societies avoid the burden of added unemployment.

> Change creates new opportunities

IMPORTANT KNOWLEDGE

Decision points

In order to respond to change it must be clear when a change has happened. It is helpful to establish what a change looks like, when it can be determined that a change has happened and what the response to the change may be.

> Change causes a need for decision, action and adaptation

In industry a change that affects the bottom-line and stakeholder confidence usually triggers decisions, actions and adaptation. The examples below look at five change scenarios, the actions they may cause and the relevant decision points.

Change:	Cloning attempt in laboratory keeps failing
Actions:	Terminate or continue the project
Decision point:	When do the resources spent on this effort exceed the projected profit derived from this project?
Reaching the decision point:	After a cost–benefit analysis

Change:	Upper management revises product portfolio
Actions:	Terminate or maintain research groups
Decision point:	Is strategy viable, competitive?
Reaching the decision point:	After market analysis

Change:	New product fails in Phase III clinical trials
Actions:	Terminate project or tweak lead for re-entry into clinical trials
Decision point	Severity of failure, cost–benefit analysis of re-entry
Reaching the decision point:	After investigation, analysis of risk and cost–benefit analysis

Change:	Consumer wins expensive lawsuit against company regarding product
Actions:	Withdraw or leave product on market
Decision point:	Severity of lawsuit, potential for new lawsuits, ethics
Reaching the decision point:	After investigation of economical implications of all outcomes and alignment with company ethics

Change:	Manufacturing plant burns to the ground
Actions:	Rebuild or close plant
Decision point:	Most profitable solution
Reaching the decision point:	After cost–benefit analysis

Market forces

Market forces are some of the most uncontrollable factors of change. They are governed by a number of events and can be influenced significantly in ways determined by cultures, economies and politics. Changes can be prepared for by understanding possible cause and effects of some of the hitherto known scenarios.

> Change can be anticipated and coping-strategies developed

In the Western world, improvement, innovation, medical treatment, quality of life and economy are in focus. Sometimes the goals in these areas clash with those of other cultures and economies and lead to tension and distraction from the goals. World crises, economic and trade disputes, sanctions, wars and natural catastrophes take their toll on stability and the goals all nations try to achieve. These crises affect both the local and global economy. Companies are most vulnerable to changes within their local area. International companies with locations in many countries may be able to escape some overall effects due to their ability to move production and people from one location to another.

Consumer confidence is essential for market optimism. Creating consumer confidence demands effective marketing and supply support. Loss of consumer confidence may cause stock prices and sales to drop. It can be very difficult to re-create consumer confidence after a loss. Successful companies appreciate the value of continuous consumer opinion, surveillance and marketing efforts, and the adaptation to changing demands.

A prospering environment with a healthy economy makes it easy for companies to grow and to adopt an attitude of high profile promises and risk-taking. The longer the environment stays healthy the more consolidated and less volatile the economic perception and foundation can become, even through times of change. In contrast, environments with high economic volatility are more likely to be influenced heavily by change.

Politics play a tremendous role in the livelihood of societies. Laws can be governed to restrict or promote growth. Ideas for the future can attract or discourage new partners. Jobs can be eliminated or retained based on national and international priorities. Most importantly, politicians are involved in advertising messages of adaptation to both the environment and workforce through their actions.

Company reorganization

Company reorganization is a term many employees fear. Reorganization can be a response to changes such as altered resources, goals, market projections and/or stakeholder demands, among other reasons. When a company reorganizes, unexpected changes are likely to happen. Lay-offs and redistribution of employees are common results as are changes in reporting structures and group compositions. Projects may be terminated, maintained or initiated. Old projects and groups may be renamed or merged, for example. Because reorganizations often are very time-consuming events, it is wise to be knowledgeable about the most current reorganization in a company and whether or not a new reorganization is likely to occur shortly.

> Expect reorganization and anticipate the effects

Company reorganization management

Accepting that change is likely to happen becomes the first step in preparing for change. The second step is anticipating the nature of the change and deciding how to respond to it. When the change occurs action must be taken to adapt and, if possible, help others to adapt. Three different responses to company reorganization are presented in the scenarios below.

Change:	Company reorganization
Nature of change:	Lay-off
Response:	Obtain good references, rethink career, find new job. Ask for severance. Ask for transition management. Leave on good terms.

Change:	Company reorganization
Nature of change:	Project change
Response:	Adapt

Change:	Company reorganization
Nature of change:	Responsibility reallocation
Response:	Evaluate if new job load and placement is consistent with own priorities. If this is the case, adapt. If not, find a new job.

Anecdote: Company reorganization

Will, a scientist, was going over new protocols in the laboratory with his technician, Linda, when Will and Linda's supervisor, Mark, entered, seemingly stressed. "I need to speak with Linda immediately," he said.

Linda went with Mark to his office. Mark said: "Linda, I have really enjoyed working with you. You probably know that the company is undergoing reorganization and I have been asked to thank you for your services and escort you from the building within the next 30 minutes. Please go to your desk and pick up your personal belongings. Leave all notebooks and company material with me, including your badge. Your computer access has been canceled so do not worry about logging in. Do you have any questions?"

Linda was in shock. Although she was able to thank Mark for the information, go to her desk and quietly collect her things before Mark escorted her out of the building, took her keys and left her outside, she had no idea what to do next.

Mark went back to the laboratory and called Will to his office. "Will, many people have been laid off today, and so has Linda. You will have to handle her workload. We also need to work more efficiently because we are under increasing pressure from the top to produce more. You may have to stop changing and optimizing the protocols and just stay with a more production-oriented procedure."

Will was surprised. He had done everything he could during the past three years to make the procedures state of the art. He believed that if the company was to stay competitive this work would need to continue. In other words, he disagreed with the decision and started complaining about this and his increased workload to his co-workers. Due to this reluctance to adapt Will was laid off during a second phase of reorganization.

Linda's lay-off situation is a common one. The way a lay-off is performed is often very surprising, carried out without notice, and seems impersonal and cruel. It is normal to be in shock and not to know what to do afterwards. Help is sometimes given to good employees in the form of career counseling, transition management and maybe a severance package. Both Will and Mark would most likely also be able to help Linda understand the practical and non-personal nature of the lay-off if the reason for Linda's lay-off was entirely business-related.

Will's situation is also very common. Employees can become disgruntled with changes with which they do not agree and they risk ending up on the "short-list." It is advisable to display compliance with change. It is easier to secure a new job while employed than when unemployed.

Summary of key skills for success

To be flexible you need to:

1. Develop change-coping skills
2. Adapt quickly to change
3. See change as an opportunity for improvement
4. Understand the necessity of change
5. Display non-possessive behavior
6. Stay neutral or positive towards change
7. Avoid taking things personally unless there is a reason to do so
8. Help others be flexible
9. Look into what is next

Mindset assessment: Flexibility

To assess whether or not you are flexible in your job situation you may take a look at how you answer the following set of questions:

What happens if you are asked to do any of the following actions?

1. Terminate your project
2. Take on new responsibilities
3. Work in a new group
4. Work on a new project

5. Work under a new supervisor
6. Change deadlines
7. Change deliverables
8. Leave your job

You can choose to react to the above changes with anger and frustration. What is happening may not seem fair to you and you may have another opinion about what is necessary. You may see the change as a personal insult and may feel hurt. You may even complain about the decisions and try to circumvent or change them. This attitude is not flexible.

Alternatively, you can choose to display a neutral or positive attitude. You may decide to stay pro-active and see if the changes can lead to a better and improved situation for you and the company. Maybe you believe changes are necessary and you take the changes as they come, finding ways to adjust and adapt. It may even be exciting to challenge yourself in new ways and you may be grateful for the opportunity. This attitude is flexible.

Once a change in project or a lay-off has happened to you, you are well aware of the loss it encompasses and also of the fact that you can survive the change. It may take some time to cope, but the less personally you experience these events, the sooner you are going to be able to adapt to your new situation.

> Put things in perspective:
> losing a job is not the same as losing life,
> and something great could come out of it

Chapter

10

CREATIVITY AND OUT-OF-THE-BOX THINKING

Bringing new, untraditional thoughts, ideas and technology to an area can be characterized as an out-of-the-box contribution. Thinking out-of-the-box is a creative behavior that carries substantial risk for both success and failure. Great discoveries and new inventions are nurtured in creative environments.

SCENARIO AND IMPACT

In all companies there is an interest and need to discover, invent, patent, manufacture, market and launch products.

> A company's ability to expand its intellectual property portfolio and to improve is dependent on inventions

In order to create inventions, one must know when something is new. Employees must, therefore, keep up to date with new technology, literature and patents within their areas of interest (Chapter 13). Interaction with other people and other fields challenges the integrative abilities of an investigative and analytical mind. Having time to experiment outside set deliverables is necessary if new inventions are a high priority.

> Scientists invent during recess: "play-time"

Most scientists will be excited to have such an opportunity and feel they are working in a creative environment. It can be quite a contagious and positive influence that stimulates more discoveries.

IMPORTANT KNOWLEDGE

Context understanding

Making inventions requires attention to detail and understanding of how the details will fit into the "big picture." Some inventions are simply based on thoughts established in one context and used in another. Therefore, "borrowing" thinking from other fields and contexts and being able to understand the contexts are important. An example of this is mimicking, e.g., bio-mimicking, where nature's creations are imitated as various inventions.

> Context-oriented thinking helps to set goals beyond details

Inter- and multi-disciplinary thinking

Integration of different disciplines into a larger study creates opportunities for multi-faceted innovation. A great example is the promise of genomics (see Chapter 13), which employs and integrates data at many levels (DNA, gene, gene function, protein, metabolite and regulatory elements, etc.) in order to understand whole organisms and their states of health and behaviors.

> Using information from multiple disciplines, traditions and cultures increases opportunity for discovery

There are various ways to create multi-disciplinary thinking. One way is to bring people of different educational and cultural backgrounds together to analyze the same problem. Their different frames of mind constitute a large resource of opportunity.

Cultural and educational diversity is a resource for innovation

Innovation and entrepreneurship

In companies where innovation is of high priority (most companies, especially start-up and companies with a large research component), people with entrepreneurial qualities must be hired. Entrepreneurs are risk-takers. They believe passionately in their inventions and they persevere through times of adversity to bring ideas from product to market. Entrepreneurial thinking is nurtured by unmet needs and problems within environments where a solution is essential and which provide access to the required tools for invention.

Entrepreneurs solve problems in untraditional manners, thereby creating inventions

Metrics and creativity

As companies grow and become successful, so does the need for organization and measurement of success. Often, so-called metrics are introduced. Metrics define certain amounts of deliverable units. These are usually established for each work group. For example, a production group may have to produce 200 liters of vaccine in a year, an intellectual property group may have to file 10 patents per month, or a research group may have to test two new methods for optimization of production per week.

Metrics and creativity can be synergists at the optimal level and antagonists if the metrics are time-constraining

When the pressure to produce leaves time to invent ways to improve production, the balance between metrics and creativity is optimal. This means employees have time to optimize and improve on what they are doing. On the contrary, if metrics are very demanding, employees may have no time at all to improve on what they are doing because all of their time is used just to maintain the metrics which are required of them.

Anecdote: Innovation obstacles

Christine was an enthusiastic entrepreneur. In her previous job she had managed to obtain multiple patents and produced several publications while she was able to meet her regular deliverables. She was a highly valued employee.

Her husband was relocated to a new area far away from her job. She decided to follow him to the same area and accepted a job in a production facility.

The new job turned out to be far less entrepreneurial than the old job. It did not take long before Christine realized that the pressure to produce deliverables was increasing and that she missed inventing. She needed time to "play" in the laboratory, time to keep up with literature and time to visit universities and academics to hear about their latest ideas in order to thrive.

Christine got a large end-of-the-year bonus. She had exceeded her expectations and was considered an outstanding employee. Despite the recognition, Christine resigned and accepted a position at a small start-up company. She went back to experiment and invent. She was thriving again and it did not take long before she re-established ties with her former employer and started to collaborate.

Christine's situation is not unusual. She had experience in two very different companies; one of them offered a sufficient level of freedom for innovation (in order to meet her needs) whereas the other did not. Different companies place different emphasis on innovation. For example, some companies will specify that an employee may use 10% of the company's time for experimentation, whereas others may not specify any time for this endeavor. There are many variations and it is wise to do some research to see if the innovation environment is sufficient (for personal needs). In Christine's case she seemed to have gone from one industry extreme to the other and finally decided to return to an environment resembling her initial industry setting.

Summary of key skills for success

In order to be creative and think out-of-the-box you need to:

1. Have time to think
2. Be able to see the big picture (context)
3. Pay attention to detail
4. Be willing to integrate knowledge from different disciplines
5. Be willing to integrate knowledge from different cultures
6. Interact with other people
7. Seek information from many sources
8. See problems as opportunities
9. Appreciate improvements
10. Take risks

Mindset assessment: Entrepreneurship

Are you able to think out-of-the-box? The following series of questions and answers will help you assess whether you currently enjoy entrepreneurial thinking or if you are operating to produce metrics. Mindset 1 is metrics-oriented and Mindset 2 is entrepreneurial. If you find yourself entirely aligned with Mindset 1, you are most likely to thrive in a contract environment; if you identify with Mindset 2, you may be most successful in a research group and/or start-up company. If you find yourself flexible you are likely to be doing a great job no matter where you are.

Question	Mindset 1	Mindset 2
How do you approach a problem?	I ask somebody about a predetermined solution	I collect a lot of information to see different points of view before I suggest a solution
Do you like to have extra time to experiment on your own?	No, I like it when my tasks are absolutely clear, no surprises, please	Yes, I like to have time to try some of my new ideas out in the laboratory

(Continued)

Question	Mindset 1	Mindset 2
Are you a detail-oriented person?	Yes, extremely detail-oriented	Somewhat, but I also have a good feeling for the larger prospects and context
Do you like to be involved in innovation?	Not particularly	Yes, that is what I prefer
Are you able to summarize and simplify information from many different sources?	No, I lose track of the details if I have too much information	Yes, I like to do so rather than focus on all the details
Do you like to try new ways to solve old problems?	I do not see reasons to fix things that are already working well	Yes

Chapter

11

AMBIGUITY MANAGEMENT

Ambiguity is a challenge to the human mind because a situation of uncertainty that demands a solution has been created. Often it is not easy to create a solution and make the necessary choices, hence the ambiguity.

SCENARIO AND IMPACT

In most companies a high number of situations can be expected, which can be classified as ambiguous.

> Ambiguity is a function of knowledge and decision-making

Ambiguity simply appears because it is neither possible to know everything at the time of decision-making nor at all. Sometimes it is necessary to make a choice, take a risk and hope that the selection was the best one possible.

Uncertainty and ambiguity can be resolved, but also can be employed as tools to create highly competitive solutions to problems. The more discussion, critique and resolution a certain idea or product has faced during the development phase, the more likely it is to withstand public scrutiny once it enters the market place.

> Creation and management of ambiguity during product development can enhance product viability

How people manage ambiguity determines whether or not it is advantageous or disadvantageous for them to encounter it. Some people find it exciting and thrive in a trial and error environment with a high level of risk-taking. Others are frustrated by the constant challenges and are more likely to thrive in a predictable setting with less decision-making.

Important knowledge

Causes of ambiguity

There are certain scenarios likely to induce uncertainty in the decision-making process. If there is a lack of one or part of the three essential components: (1) time, (2) resources (knowledge, money, workforce) and (3) quality, the so-called "right" decision can be difficult to make. Often, alternative plans are made, which rearrange these three components in order to create multiple options and visualize the best possible decision.

> Lack of time, resources and/or quality creates uncertainty

- **Time.** When there is not enough time to make a decision, a "premature" decision must be made. This is a frequent occurrence in companies. Using time means using other resources too, therefore creating economic consequences. There is a fine balance between using enough time (to avoid making a major mistake) and too much time (to attempt to identify the best possible option) when making a decision.
- **Resources (knowledge, money, workforce).** Not knowing enough is a severe impediment to informed decision-making. In companies this is often the case. Due to the nature of the work, companies try to outdo their competition by inventing new and better products, methods and procedures as fast as possible. This means entering into hitherto unknown territory where having enough knowledge about every aspect is an unrealistic expectation. The speed with which a decision is desired

is also a restriction on the amount of knowledge that can be obtained within a given time-frame. Lack of money means lack of experiments, both theoretical and practical. This may lead to lack of production of data and products, which is intertwined with a lack of knowledge and adequate workforce.

- **Quality.** Compromising quality is a serious issue. Nevertheless, this is an often-used route to achieve lower costs and free up time and other resources. The quality of the product needs to meet the product specifications, not more, not less.
- **Time, resources and quality.** There is a clear interdependence between the three components of time, resources and quality. This means that if for some reason less time is given (pressure to increase production, production problems), more resources must be allocated to production or quality will decrease. Likewise, if quality has decreased, more resources and possibly more time must be given to regain the necessary specifications of quality. Figure 11.1 shows the three components.

Figure 11.1 The interdependence of time, resources and quality. The triangle represents the connection between the three elements. If the length of one side (symbolizing amount of one element) is altered, one other side or both other sides (amount of second and/or third element) must be adjusted.

The most frustrating lack of resources for many employees is the lack of exact directions. If managers are not sure of what the end product looks like, it may take many attempts (and much time) before enough knowledge, less uncertainty and firm procedures are created. The best advice is to communicate frequently with supervisors concerning the final deliverable to make sure the right approach is taken.

> Ask for feedback frequently until the end-product specifications are clear

Decision-making under constraint

There is a certain level of risk associated with making a decision within constraints. To minimize risk, an employee must know and understand the exact priorities of the company. Information about the non-negotiable bottom-line and what is left to negotiate must be obtained. Which is most important, an exact and strictly obeyed timeline, more knowledge, spending less money, using a smaller workforce, the highest quality or any combinations and variations of these components?

> Learn about the priorities and what is negotiable

Psychological effects of ambiguity

How employees deal with ambiguity has great impact on the work environment. Not knowing whether success or failure will be the outcome can be very stressful. People who have patience to wait for an answer and people who are flexible are likely to respond most appropriately. Flexible people with high self-esteem know that it is human to fail and that research means to "search again – re-search." Also, they tend not to take a failure personally (unless there is a reason to do so) and, therefore, they may be willing to take risks in a high-ambiguity environment. At the given time they will make the decision that is needed based on the available information and resources.

> Seek to make the best decision
> with the available resources

Anecdote: Time, quality or cost?

Rick was working on expanding the production of a new promising instrument. The expansion Rick was planning was a result of a request from upper management. Rick was told to double the production within six weeks "at any cost" because the company had received a large order from a prominent customer.

Rick needed to perform a series of pilot experiments to calibrate and adjust the instrument before the expansion could take place. He would not be able to manage this task by himself with the currently provided resources. He would need three more engineers in the group in order for him to meet the goal of doubling the production within six weeks.

The management team asked for a meeting with Rick. "Unfortunately, your plan is not viable to the company," he was told. The cost to hire the three extra engineers would compromise the profit. Rick would have to perform the task without the extra help.

Rick was surprised by this answer. He originally thought that cost was not an issue. Management seemed to have some ambiguity about this; at least the message had been somewhat vague. Rick knew he would have to anticipate more ambiguity and revised his plan by compromising different factors and creating different options. He presented the following three ideas to the management team.

Perform task within six weeks and without three engineers. This can be done by:

1. Using more time (compromise timeline of delivery to consumer)
2. Avoiding pilot experiments (compromise on quality and capacity of product)
3. Paying for overtime and performing pilot experiment (compromise on cost of project)

Rick preferred option 1 or 3. He did not like to compromise on quality and he preferred perfection in his work. Rick had worked long enough in the company to understand that his job was to suggest various options for performing the task so that the ambiguity could be resolved, not to make the decision.

The company chose option 2 because the general opinion was that the instruments could be adjusted at a later time after they had been delivered to the customer. Rick now understood that the highest priority for the management team turned out to be an on-time delivery in a cost-effective manner. He started the work to change the product specifications accordingly.

Summary of key skills for success

To manage ambiguity successfully you need to be able to:

1. Understand that sometimes there is not enough time
2. Understand that sometimes there are not enough resources
3. Understand that sometimes one must settle for less than perfection
4. Ambiguity can be a driver for competitiveness and innovation
5. Time, resources and quality are interdependent

6. The end-product is not always clearly defined
7. Patiently assist in defining and documenting the end-product
8. Learn and understand company priorities
9. Do the best you can with the resources you have

Mindset assessment: Ambiguity management

Are you able to work well in a high-level-ambiguity environment? The following statements will help you determine whether you can improve your ambiguity management skills. If you are in agreement with the statements you are most likely doing well.

1. I have high self-esteem and I am flexible
2. I think ambiguity creates an exciting challenge
3. I ask many questions myself
4. I am comfortable making decisions based on limited information if necessary
5. I will seek to understand the priorities before I make a decision as long as I have time and resources to do so
6. I am comfortable with my decisions and I document why I made them
7. I take responsibility for my decisions
8. I seek to succeed, but I am not afraid to fail now and then
9. I always try to do my best
10. Management needs my data and feedback in order to envision the end-product clearly
11. Perfection is great, but not at any cost
12. I am happy when I achieve exactly what is necessary and spend less rather than more resources
13. I like to invent and optimize, but understand that it comes at a price and must be managed
14. I help my co-workers deal with ambiguity when they become stressed

Chapter

12

INTELLECTUAL PROPERTY

Intellectual property is a term that relates to "possessions" created by the human mind. These "possessions" can take form as ideas and the products of such ideas (works, performances, inventions, discoveries, designs, trademarks, commercial names and designations). Because their creation usually requires an investment (salaries, materials), it is common practice to protect the "possessions" (and investments).

SCENARIO AND IMPACT

A company must invest large sums of money in research and clinical trials to create a new medicine. In order to stay viable, it is imperative for the company that it owns the right to exclude others from making the same medication. Companies with a large portfolio of intellectual property have an extra asset, in addition to the more tangible resources of a certain product or service technology. There are a variety of forms of intellectual property. The main types include copyrights, trademarks, patents and trade secrets.

- **Copyright.** Protects original works fixed in any tangible expression (books, art, music, movies, recordings, etc.). *Longevity*: Lifetime of author plus 70 years (for an individual), 95 years from the first publication or 120 years from the date of its creation, whichever is less (for works by corporations).

- **Trademark.** Protects symbols, words, slogans, pictures, designs, etc., which are used to identify and distinguish goods. *Longevity*: 10 years renewable, and therefore potentially forever if properly maintained.
- **Patent.** Protects the owner's right to exclude others from making, using, selling, offering to sell and importing the invention. *Longevity*: 20 years from filing date.
- **Trade secret.** Protects the owner's right to prevent others from disclosing the secret or misappropriating the secret (formulas, technology, instruments, lists, know-how, etc.). *Longevity*: Potentially forever if properly protected.

> Intellectual property rights exclude others from using the property

Because intellectual property can be used to limit use by others it may seem that there is a negative impact on exchange and sharing. However, companies often negotiate license agreements, which facilitate sharing of rights between the intellectual property owner and others.

Ignoring others' intellectual property may lead to acts of infringement. Infringement is an unlawful use of intellectual property. Consequences of infringement depend on legal action from the intellectual property owner.

IMPORTANT KNOWLEDGE

Confidentiality agreements

Company employees have a duty to keep certain company information confidential. The reason a company wants employees to honor confidentiality is that information could be related to intellectual property or to prospects of obtaining intellectual property. It is not always easy to know whether or not this is the case, and that is why a general attitude of confidentiality is the easiest to apply. The implications for making an invention known to the public (people not bound by a confidentiality agreement) may include the elimination of patent rights for the disclosed invention.

> Be mindful of what to say
> to whom to say it
> and where to say it

It is not only a good rule to keep certain information confidential but also a legal requirement if a confidentiality agreement has been signed. Many agreements have confidentiality provisions. Examples include confidential disclosure agreements (CDA) and agreements relating to employment, consulting and collaboration.

Confidentiality disclosure agreement statements: Example of contents

1. Persons authorized to access the confidential information
2. Type of information to be confidential and duration of confidentiality
3. Ownership of intellectual property developed using the confidential information
4. Consequences of violating the agreement
5. Procedures for terminating the CDA and consequences of termination

A CDA can be made between a number of parties, such as an employee and the employee's company (in an employment contract), the company and its collaborators (academic or commercial), or the company and its consultants, to mention some examples. The information can only be shared with people included in the agreement, neither with family members, friends, or anybody else.

Do not share confidential information with anybody (including family members) unless they are included in the confidentiality agreement

In order to know what to share or what not to share, the definition of what constitutes confidential information must be clear. Confidential information typically means all information, materials and things related to the company's business or technology, so it is highly inclusive. General information, such as any information that has already appeared in the public domain (articles, web pages, etc.) is not confidential information. Examples of information often considered to be confidential include details concerning company projects and the size and type of the company teams.

Published information is non-confidential and not patentable

The agreement normally specifies the period of time during which information must be protected as confidential. Often three- to five-year terms are required. The period may begin as soon as a party receives confidential information and the obligation to protect confidentiality may continue after the formal expiration of the agreement.

> Confidentiality is often required after the duration of the agreement

During the term of an agreement, for example an employment agreement, intellectual property may be developed. The agreement indicates who will own the intellectual property. Employees often assign all rights to the employer at the onset of their employment.

Other relevant provisions include: The need to disclose the existence of a new invention to the other party in the agreement, as well as which party will own the intellectual property and under which conditions, penalties that will apply if provisions are breached, the jurisdiction whose laws will be used to interpret the agreement, the terms of the agreement, and requirements for terminating the agreement.

Materials and technology transfer agreements

Other types of documents exist which focus on sharing of materials and technology. In addition to specific statements about materials and technology transfer, these documents usually contain statements similar to those just described for confidentiality agreements.

Material and technology transfer agreement statements: Example of contents

1. Persons, materials, technology included in the agreement
2. Type of confidential information, technology and duration of confidentiality
3. Type of research that can be performed with the materials or technology
4. Owners of intellectual property developed using the materials or technology during the agreement
5. Special conditions (storage, handling and shipping)

6. Return of materials
7. Consequences if the agreement is violated
8. Terms for termination of the agreement
9. The beginning and end of the agreement

Inventions and invention disclosures

When an employee thinks he or she may have invented something, the first step is to properly disclose the potential invention. Employees must file an invention disclosure form (IDF) to an intellectual property committee for review. This committee oversees the intellectual property status within the company. An invention disclosure form contains a description of the invention, the time of invention, the name of the inventor, competing works and potential use of the invention. The committee will perform an analysis to assess whether or not an invention has been made. This analysis includes searching of academic, government and commercial databases (including the patent databases). All **prior art** – identical, similar, partial or competing inventions – must be identified.

Classification of intellectual property

The invention needs to be assessed in terms of relevance from a business goals perspective. The prospective intellectual property needs to be classified to the best form of protection: Copyright, trademark, patent or trade secret. Often a single invention can be protected by more than one type of intellectual property. For example, a new technique is invented that will be patented. The name of the technique will be protected as a trademark, the document describing the technique will be protected under the copyright laws, and the technique itself will be patented.

Whether or not the invention justifies a patent application or should be kept as a trade secret depends on the company's immediate need, the status of the competition and market analysis, and the cost of obtaining patent protection, which can be quite expensive.

Patents

In order to obtain a patent for protecting an invention, certain conventions must be followed. First, the idea must fulfill the requirements for patentability and remain undisclosed to the public (although see below for the US exception to this). Second, the idea must be documented. Third, a patent application describing the idea using legalistic language must be submitted. Finally, correspondence between the patent granting organization and the potential patent owner must take place in order to define the boundaries of the intellectual property.

Anything made by man is potentially patentable subject matter as long as it is a new and useful process, machine, manufacture, composition of matter, or any new and useful improvement thereof. Patentable subject matter must have **utility**, be **novel**, **non-obvious** and **enabled** and have a **written description** (US guidelines). A simplified explanation of each of these concepts is given below.

- **Utility.** Be operative and have substantial and credible utility with feasible use.
- **Novelty.** Not already available to the public, not already known or used in the United States, or patented or described elsewhere.
- **Non-obviousness.** Not possible to be readily deduced based on publicly available information.
- **Enablement.** Application must describe how to make and use the invention.
- **Written description.** The invention must be described.

The language in a patent application is a combination of scientific and legal language. It takes patience and dedication to learn to read and write this language. This is the reason companies employ patent attorneys and patent agents for such tasks. Working with the attorneys/agents to transpose the invention into legal claims is a worthwhile and exciting experience.

> Learning to write legal claims
> is like learning a new language

Once the patent application has been written and submitted it will be evaluated by the central patenting authority in the country where it is filed. Claims will be scrutinized and may be rejected. It sometimes takes many years to negotiate the claim language that the examining authority will consider acceptable. This can be an unexpected and somewhat stressful experience for first-timers.

Assembling an intellectual property portfolio is a global endeavor. This constitutes a challenge because of the difference in laws, cultures and each country's level of compliance with international rules. Although a series of streamlining international agreements has been signed by most countries, differences among countries remain in the rules for obtaining and maintaining patents. Important examples of these differences include the rules about what can be patented, the effect of public disclosure on the right to obtain a patent, and the rules about who wins if two companies file patents for the same invention:

1. A European patent is granted to the person who is the first to file the invention (deemed by filing date), whereas a US patent is granted to the person who is the first to invent (deemed by documentation records).
2. In the United States an inventor may file a patent up to one year after public disclosure. Everywhere else public disclosure prohibits granting of a patent.

Publications

There are several ways to "publicize" an invention. Written publication takes place if the invention is described in a meeting **abstract**, on a **poster**, in a scientific **journal**, as a **press release**, in an **annual report**, on the **Web**, or in **any other way expressed in print**. Oral publication is any **speech** at **public conventions, meetings**, at the **dinner table** or in **any setting** including people who are not bound by confidentiality agreements, **in which the invention is orally described**.

> Making an invention known by oral or written means to members of the public who are not bound by a confidentiality agreement is a publication of the invention

Companies take great care to train their employees how to honor their confidentiality agreements, so they know how to avoid making potential inventions known to the public before they are properly protected. Especially at large scientific meetings, companies may present data while withholding certain critical information.

Permission to publish

Most companies have rules about obtaining permission to publish. It can take a while to obtain this permission and it may not always be granted (hence the need to obtain permission). The following description of an application form for permission to publish serves as an example. Typically, permission needs to be granted by several people who are able to assess the consequences of publication. Such people may include scientists, most likely the nearest supervisor, a research executive, a legal counsel representative and maybe the CEO of the company. The form contains information relating to the work and its publication. Figure 12.1 suggests some of the issues that may need to be addressed and the signatures that may need to be collected in order for the publication event to be in compliance with the company guidelines.

Application Form: Permission to Publish

Name ______________________________

Group ______________________________

Supervisor ______________________________

Date of expected publication ______________________________

Publication medium (talk, poster, paper, book, other) ______________________________

Materials to be published (attach exact materials) ______________________________

Have these materials been published before (if yes, when) ______________________________

Has a patent application been filed (if yes, when) ______________________________

Will (additional) patent application(s) be filed (if yes, when) ______________________________

Applicant: signature and date ______________________________

Approval, Supervisor: signature and date ______________________________

Approval, VP of Research: signature and date ______________________________

Approval, Legal Counsel: signature and date ______________________________

Approval, CEO: signature and date ______________________________

Figure 12.1 Application to publish

After obtaining approval the exact information that was approved (nothing else) is released for publication.

> Only publish exactly what has been approved for publication

Once a paper or a patent application is published or a patent is granted, all of the content can be considered contained within the public domain if the publication databases are accessible to the general public. It is no longer necessary to gain internal approval for additional publication events.

Inventor/authorship of patents and publications

As mentioned, company employees who are inventors and authors on a patent application normally assign all ownership for patents to the company for which they work. They do this at the start of their employment. In contrast, employees at universities often share the ownership with the university for which they work. In both cases, it is essential to understand who is considered an inventor and should thus be listed on a patent application.

> A person who conceives an invention is an inventor

Unlike authorship on scientific papers (see Chapter 6) inventorship in a patent application is determined by specific legal standards. An inventor is the person who conceived the idea or part of the idea for the invention as it is claimed in the application. If the idea was conceived at a group meeting, the members of the group may all be considered inventors. Excluding an inventor can be fatal for a patent if the patent is later in dispute, because intentional exclusion of an inventor can invalidate the patent.

> Include all inventors on the patent application

An inventor is *not necessarily* the person who carries out the work to produce the invention unless this is the person who also had the idea for the invention. What this means is no matter how much work it takes to verify the invention, the work does not always justify a status as inventor on a patent application.

> A person who makes the invention work
> is not necessarily an inventor

Inventors share the same rank on a patent application. Therefore, only the appearance of names and not the order in which the names appear has significance. This is different from the common practice in journal articles (see Chapter 6). However, the patent will often be referred to by the name of the first inventor.

Documentation

For all rules, regulations, patent and publication purposes documentation is necessary. Documentation is performed individually, preferably each day, by employees as entries in a personal notebook owned by the company. The notebook must be contained within the company. Companies usually have their own specific guidelines. The notebook pages may look something like Figure 12.2.

CONFIDENTIAL

Title:____________________ Project:____________________ Book No:______

your project here

Signature____ Date________

Witnessed & understood by

Signature____ Date________

Figure 12.2 Example of a notebook page

There are common rules frequently used for documentation. For example, all entries should be made with non-erasable ink. If corrections are needed, these must be made by crossing over old entries (so these still are legible) and making new entries. Experiments should be legible and easy to follow. As a rule, a person who does not know the experiment should be able to pick up the notebook and understand the entries. All pages need to be signed and dated. This is especially important in case an invention is recorded because it is essential to know the exact date the invention was conceived. For the exact same purpose the notebook must be witnessed, understood and signed by a colleague (often from another group or field of work).

Anecdote: Patent ownership

Harry was currently testing a new hypothesis of his and was on the verge of an interesting discovery.

Mike and Harry went to lunch together. They found a quiet place in the restaurant, well assured that nobody was going to listen in on the conversation. Harry kept the information very general. After all, this was a public place, so he was not going to take the chance that anybody from other companies would be able to hear what he was saying. He knew he had to be careful in order to protect the potential invention and intellectual property.

Mike listened to Harry and his ideas. After returning from the restaurant, Harry wrote an invention disclosure form, signed it and submitted it to the company's legal office.

The following week was crucial for the verification of the hypothesis. The experiments confirmed the theory and it was clear that Harry had a breakthrough; he had invented a new technology. Upper management was notified about this and Harry was asked to provide information for a patent application, which would be written by the company lawyers.

This was the first time Harry had interacted so closely with the lawyers. It was amazing how differently they saw things, especially how they wrote about them. It was like using a totally different language. After two years of back-and-forth submission and negotiation with the official patent office, the patent was issued. In the meantime, Harry and Mike had been spending time discussing the invention thoroughly and they had been writing a publication together. The publication would be submitted after the patent was granted.

Mike realized that his name was not on the patent but was on the publication. His first reaction was disappointment. He asked his supervisor for an explanation. Mike's supervisor understood the problem. It was common and happened all the time. He said: "An author of an invention is the person who has the idea. This is also the person who submits the invention disclosure form. It is important that all inventors are named on the invention disclosure form. But if it's not your idea and you did not contribute to the idea, merely listened, you are not an inventor. You may become a co-author on a journal publication, if you continue to listen, get involved, contribute academically or experimentally to further results and give feedback, and if Harry is planning to write one." Mike understood the difference. He was proud that Harry had asked him to be a co-author on the journal publication.

Summary of key skills for success

To develop intellectual property you need to:

1. Document findings in your notebook
2. Sign and date your notebook
3. Get a colleague to witness, sign and date your notebook
4. Write and submit an invention disclosure form if you think you have made an invention
5. Write and submit a patent application
6. Negotiate claims

To protect intellectual property you need to:

1. Keep your findings confidential
2. Record your findings in appropriate media in a timely manner
3. Include all prior art when evaluating potential new intellectual property
4. Disclose your findings to appropriate committees in a timely manner
5. Include all inventors on the patent application
6. If you want to publish your findings, obtain permission to do so
7. Publish your findings only *after* the patent application has been filed

Mindset assessment: Intellectual property

Are you knowledgeable about the process of obtaining intellectual property, how to protect the intellectual property and what it means for the company at which you work? The following statements will help you assess your current mindset. If you are in agreement with the statements you most likely understand the most important concepts and are able to act appropriately.

1. I keep a notebook
2. I sign and date my notebook regularly
3. One of my colleagues witnesses, signs and dates my notebook regularly
4. I am conscious of when I may have made an invention

5. I study the prior art before I decide whether or not I have made an invention
6. I submit invention disclosures in a timely manner
7. I am interested in learning how to write good claims
8. I interact well with our legal counsel
9. I make sure I include all inventors on the invention disclosure form
10. I am aware that obtaining a patent can be a long process
11. Patience is required in the negotiations with the central patent office
12. I can keep a secret
13. I am very careful about what I say in public
14. I honor my confidentiality agreements
15. I request permission to publish my work before I publish it
16. I respect the decision if my request to publish is not granted
17. I am patient in terms of publishing my data
18. I know that when I publish something, it cannot be patented later (except in the United States, where I have a one-year period to file a patent application).

Chapter

13

SPECIALTY TECHNOLOGY AND KNOWLEDGE

A large set of technologies and knowledge accompany any industry. It is an insurmountable task to attempt to know it all before entering into an actual job. It is also unrealistic to cover this in a book because the technology changes rapidly and so does knowledge about the particular areas.

> The technology is constantly changing and knowledge is accumulating

This chapter provides a couple of examples of the kinds of technology and knowledge that are currently in high demand in biotechnology and pharmaceutical companies.

SCENARIO AND IMPACT

It is particularly important to stay up-to-date in technology-intensive environments such as those found in the biotechnology and pharmaceutical industries. In these environments recent discoveries have caused a rapid influx of new technology, equipment and methods. In addition, an entire field dedicated to quality of data, processes and products has developed (see Chapter 14).

> Stay informed about progress made in your profession

The reason it is important to stay knowledgeable about progress is that both employees and the company in which the employees work are more likely to remain competitive. For example, if there are new ways to develop better medicines faster, it is essential to know about it. The new knowledge may have the potential to decrease the secondary effects of a medication and provide a clear advantage to the consumer, becoming the preferred medicine. Knowledge about progress in a field can be obtained through many different means, for example:

1. Interaction with other people at scientific meetings, symposia, workshops
2. Continuous training in new technologies
3. Discussion with academic and industry colleagues
4. Reading of literature, including patents, press releases, FDA (Food and Drug Administration, see Chapter 14) approvals, etc.

If a company stays competitive, jobs are likely to be preserved. In addition, if employees are competitive, they are much more likely to increase their flexibility on the job market.

> Stay up to date – stay competitive – stay in demand

New technology, knowledge and thinking about improvements foster innovation and creativity. Because industry rarely adheres to a situation of status quo, these qualities are almost always appreciated.

IMPORTANT KNOWLEDGE

Practical experience

Practical experience in new technology is considered to be one of the best assets an employee can have. Finding a job in a company that is at the forefront technologically may be optimal but not always possible, in particular if the technology or equipment has just been developed. Fortunately, there may be other options to learn. Education and training workshops may be offered in the use of the technology after its publication. Obtaining training and/or continued education are means of staying competitive.

Keep educating yourself

Licenses

Accumulating relevant certificates and licenses is another way to stay up to date. Many companies offer employees enrollment in continued education and opportunities for various workshops and certification programs. Being a licensed professional may provide an advantage in a job market that is increasingly competitive. In the biotechnology and pharmaceutical industry, licenses and certifications are available within many of the professions, some of the most common of which are:

1. Patent liaison
2. Project manager
3. Regulatory professional
4. Safety manager
5. Operator of specialty equipment
6. Leadership certification.

High demand technology in the biotechnology and pharmaceutical industries

Genomics

A genome constitutes all of the genes in an organism, and genomics is the study of the genome. The advent of genomics has caused extensive use of "high throughput" methods. These methods make it possible to look at very large data sets in a short amount of time. In addition, multiple data sets can be integrated and sophisticated analysis can take place. There are several disciplines to master when entering this field, based on the various levels of analysis.

DNA analysis

1. Genomic DNA extraction and library production
2. Sequencing and database searching
3. Gene knock-out technology
4. Gene amplification technology.

The DNA analysis capabilities concern the ability to capture an organism's DNA into so-called genomic libraries, to characterize the captured DNA down to each base (A = adenine, C = cytosine, G = guanine, T = thymine), to understand which kind of genes these bases constitute, to shut down the genes and to increase the function of these genes.

RNA analysis

1. RNA extraction and cDNA library production
2. Sequencing and database searching
3. Gene expression profiling.

The RNA analysis techniques include the ability to capture the product of "active DNA," that is, DNA that is being expressed as transcripts, to characterize the transcripts down to base (A = adenine, C = cytosine, G = guanine, U = uracil), to understand which kind of genes these bases constitute, and to understand the level of activity of these genes at a given moment.

Protein analysis

1. Protein purification and characterization
2. Heterogeneous protein production
3. Protein modification and optimization of production.

The protein analysis techniques encompass the identification and characterization of proteins of interest, the production of the particular proteins in new hosts, the manipulation of the proteins, and ways to increase production of the proteins.

Drug discovery

The drug discovery process leads to the production of new drugs. It consists of many different interrelating parts. These are divided into five categories for reasons of simplicity and the typical work in each of these in a US environment is described: Research, Preclinical studies, Clinical studies, Regulatory approval, Launch and support.

Research

1. Research topic decision
2. Identification of target (gene, transcript, protein) associated with research topic
3. Lead identification
4. Prioritization of leads.

The first major decision is what to pursue. When making new drugs to treat a disease there are several issues to consider such as: Is there a need for the new drug? What is the market potential? What is involved in identifying a new drug? What is the cost of the final product? How long can it keep on the shelf? How can it be distributed? What is the level of competition?

Once the decision has been made to pursue making a new product, the research groups must identify which kinds of targets need studying. Often it is a matter of identifying and inhibiting certain genes, pathways or products of pathways. Depending on the target, the ways to identify this and to produce leads that inhibit the target must be devised.

Lead identification often involves a step of synthesis of a large number of compounds. These will potentially interact with the target followed by high throughput screening of the interaction between compounds and target. When several leads have been identified these must be prioritized according to their effect, their dosage response and delivery method, and the feasibility for manufacturing. An IND (Investigational New Drug application) is usually filed at this point.

Preclinical studies

1. Large scale manufacturing
2. ADME studies (see below).

Leads are characterized according to their feasibility for large scale manufacturing and formulation. Toxicological studies are initiated and the first animal studies may be conducted. Lead compounds are characterized based on *absorption*, *distribution*, *metabolism* and *excretion* of the compound. In addition, the shelf life, purity, dosage and dosage form is determined in this phase.

Clinical studies

1. Phase I
2. Phase II
3. Phase III.

Before clinical studies can start, a "First In Humans" (FIH) application must be submitted and approved. During Phase I trials the lead is tested in a few healthy human volunteers. Its safety and how well it is tolerated by healthy humans is assessed. This step constitutes the proof of concept stage. In Phase II a mix of 100–300 healthy and diseased volunteers are assessed. Finally, Phase III includes thousands of volunteers. During this phase the NDA (New Drug Application) is prepared.

Regulatory approval

1. Filing with regulatory agency.

After filing of the NDA to the FDA, the FDA will process the claims and, if approval is given, the company can move to the final phases in the drug discovery pipeline. Registration can be initiated and plans for production can be made.

Launch and support

1. Marketing and launch
2. Phase IV.

During this step the new drug is marketed and distributed to the consumers. During this time, the clinical trials, in particular Phase IV, which is a safety phase where the intake of the drug is continuously monitored, continue. This is also a good time to consolidate the intellectual property. Product line extensions are often considered.

Anecdote: Staying competitive

Jeff was in charge of new technology implementation. Basically, he supported all of the groups in molecular biology. He would investigate what was new, go to meetings and talk to other companies and universities about the newest innovations. He would also make sure he had the latest training, and for that purpose he maintained friendships with various technical specialists world-wide.

Despite Jeff's great work, things took a turn for the worse. Administration had been corrupted for a while and there had been a severe outflow of money. This caused a decrease in the revenue large enough to destabilize consumer confidence. The board was fired and "heads were rolling." Jeff was among the people who were laid off in the first round because it was assumed that no new technology would be needed for a while.

Many people were shocked about Jeff's dismissal and thought it was unfair. At the same time there was relief among those who got to stay. Jeff was surprised but not worried. He had so many contacts and was used to interacting professionally with others. Most of all, he was one of the best qualified within his field. Other companies had tried to head-hunt him several times. He had not accepted any offers then, but kept in touch and saved the contact information. Again, he was confident about his own situation. Jeff knew that there would always be somebody who would be looking for someone like him as long as he was up-to-date with all of the new technology. Jeff found a good new job very fast.

Summary of key skills for success

To stay informed about specialty technology and knowledge in your field you must be able to:

1. Learn from others
2. Read newsletters, journals, press releases, technical notes, patent applications, etc.
3. Go to meetings and interact with others in your field
4. Obtain relevant training
5. Obtain practical experience
6. Obtain a license if necessary
7. Stay up to date

Mindset assessment: Technology in the biotechnology and pharmaceutical industries

Do you know which kinds of technologies are in high demand in the biotechnology and pharmaceutical industries? Are you familiar with genomics and drug discovery? The following questions and answers can help you assess your knowledge. If you are able to answer the questions as indicated you are most likely knowledgeable about parts of the big picture in these areas.

Question	Answer
What is a genome?	The gene complement in an organism
What is DNA?	Deoxyribonucleic acid, DNA contains genes
What is RNA?	Ribonucleic acid, mRNA is produced from active genes
What role does mRNA play?	mRNA is translated into protein
What does HTS stand for?	High Throughput Screening
Mention how HTS can be done at the DNA level.	DNA sequencing. Entire sequences, part of sequences or a particular point mutation can be identified
Mention how HTS can be done at the RNA level.	Gene expression profiling. A decrease, increase or no change in expression at given time points, at different treatments and treatment levels can be identified
Mention how HTS can be done at the protein level.	Protein profiling. A decrease, increase or no change in translation at given time points, at different treatments and treatment levels can be identified
How does one perform HTS screening?	Robotics technology is involved at most levels
How do genomics and HTS fit in with drug discovery?	Most drug discovery involves the use of large libraries of DNA, RNA and/or protein as well as lead compounds
How does a company decide whether a new drug should be pursued?	The drug needs to be commercially viable, the production needs to be feasible, and the regulatory agencies need to embrace the idea
What characterizes a good lead compound?	It interacts with the target effectively, it can be manufactured in large quantities, it can be delivered to the target, it has low or no toxicity, high efficacy, long shelf life, high purity and potency

(Continued)

Question	Answer
Which kinds of volunteers occur in the various phases of clinical trials?	Animals or other organisms (Pre-clinical phase) Healthy volunteers (Phase I) Healthy and diseased volunteers (Phases II, III, IV)
Most people do not think about Phase IV. What happens in this phase?	The effect of the drug is continuously monitored. If severe effects are identified (side effects, death) the product may be removed from the market
How do companies support an existing product?	By constant marketing, by investigating whether more applications can be found or if slight changes can lead to more applications and products (product line extension)
What is the difference between the biotechnology discovery pipeline and the pharmaceutical drug discovery pipeline?	In cases where products are not considered foods or drugs, clinical trials are not necessary.

Chapter

14

QUALITY

A product's quality is measured by its adherence to or deviation from predetermined quality standards. In other words, if a product performs as expected and is consistently produced, the product and its production are of acceptable quality. The challenge is to determine the parameters by which quality can be defined and the product quality assessed.

> Defining discrete and universal measures of quality is essential

SCENARIO AND IMPACT

In many industry environments, specific quality standards are required and overseeing that these standards are kept can be a major endeavor. This is especially true for the food and drug industry, which is highly regulated (including the pharmaceutical industry, whose regulations are the focus of this chapter). The standards in the food and drug industry are provided by specific regulatory agencies.

Different environments and project stages have different types of regulations with which to comply; together these are labeled as CGXPs, Current Good Practices.

CGXPs ensure a standard "code of conduct" in multiple practices leading to the final product

As the name implies, these practices are current and, therefore, dynamic. In the drug discovery process (see Chapter 13), relevant CGXPs include:

1. CGLPs = Current Good Laboratory Practices. Implementation of CGLPs often takes place after the initial lead identification stage and prior to the pre-clinical studies. Once implemented, CGLPs must be maintained throughout the proceeding steps in the drug discovery process.
2. CGCPs = Current Good Clinical Practices. Implementation takes place at the initiation of clinical studies (Phase I, II, III) and continues throughout these studies.
3. CGMPs = Current Good Manufacturing Practices. This set of requirements is introduced at the onset of the manufacturing process. It continues through Phase IV and throughout the lifetime of the product.

Adhering to the set standards is an expected part of the enterprise of drug discovery and production. One could say it is an indirect "contract" with the consumer. When the product is of the expected quality, the product will be released for sale.

A product produced under CGXP oversight is expected to be of a certain (approved) quality

The reasons for and impact of having a set of standards/expectations for a product and reinforcing these expectations are many:

1. Having a universal "code of conduct," the CGXPs **ensure an "agreed-to" standard** and makes standards, processes and products **transferable geographically**.
2. Having set expectations of production and products helps **avoid product batches and processes that are not up to standard**. Such batches could be contaminated or just performing to a lesser standard.
3. Avoiding contaminated batches that do not perform as expected helps **minimize the occurrence of unexpected effects** (side effects, deaths), thereby creating a **higher level of safety for the consumer** when compared to products that are not under CGXP oversight.

4. Due to the regulatory oversight there is a benefit of **minimized risk of deviations from the standards, mistakes and fraud**.
5. Minimizing the occurrence of deviations means **lessening the number of investigations necessary** to identify the cause of the deviations, thereby **lowering the cost of production**.

> A company that develops foods or drugs must adhere to required regulations; in doing so, it must develop and adhere to CGXP compliant standard operating procedures relating to processes, products and performance

Important knowledge

Quality in research vs. production

During the different stages of research and production of a new product the number and type of CGXPs differ. In the early research phases, when new drug candidates are being identified, compliance with no or very few regulations is required. The consistency of results and quality in this phase may vary and the research is highly qualitative. Later, when a drug candidate is promoted to a lead candidate and toxicology and manufacturing studies are at the onset, the first "good practices" need to be present. Once initiated, the requirement for CGXPs expands during the following stages to encompass laboratory, clinical and manufacturing procedures.

Standard operating procedures

Standard operating procedures (SOPs) are operating procedures, protocols, or instructions that have been standardized. A well-written SOP is so detailed that no misunderstanding can occur and prevents mistakes from being made unnoticed. Regulatory agencies require that SOPs are in place for all aspects of production of a drug and that they are in CGXP compliance. For example, SOPs must be in place for different actions such as how to dress in a manufacturing plant (gowning), operation of equipment, cleaning procedures, how to measure the quality of a product batch or a procedure, how to document failure to meet specifications and how to perform an emergency evacuation from the plant.

A good SOP provides detailed step-by-step instructions and leaves no room for misunderstanding of the procedure

Example: SOP for signature of notebook entry
After each entry on a new page immediately sign and date the page on which an entry was made:

1. Use a black, non-erasable ink pen.
2. Sign your name in the box labeled "Author signature."
3. Write the date in the box labeled "Author date" (month/date/year).
4. A peer from another group must also witness, understand and sign the same page in your notebook with black non-erasable ink in the box labeled "Witnessed, understood and signed by."
5. At the same time the peer must also date the page in your notebook with black non-erasable ink in the box labeled "Witnessed date."
6. You must have the new entries in your notebook signed and witnessed by a peer once per month.

Regulatory requirements and compliance

CGXPs are specified by the relevant regulatory agencies. The responsibility for implementation of the CGXPs, the quality standards and the oversight lie with the company and regulatory agencies. Companies in different countries that want to interact, collaborate or distribute and sell their products must comply with foreign regulations in addition to regulations in their country of origin.

The most relevant US regulations for drug production are briefly presented below. Regulations are found in the Code of Federal Regulations (CFR) and these are divided into Titles. Title 21 is designed for Food and Drugs and, therefore, relevant to the pharmaceutical industry. Each title is divided into parts and each part addresses different aspects of the processes needed to develop, produce, store and sell food and drugs. Similar regulations exist in many other countries. The International Conference on Harmonization of Technical Requirements for Registration of Pharmaceuticals for Human Use (ICH) has been established to integrate regulatory guidelines from Europe,

Japan and the United States. Certain areas may contain a mix of regulations, for example: GCP regulations: 21 CFR 11, 50, 54, 56, 312, 314; ICH E2A, E6; GMP regulations: 21 CFR 11, 210, 211, 820, ICHQ7A; Medical device regulations: 21 CFR 11, 50, 54, 56, 807, 812, 814, 820, 510K.

Table 14.1 displays samples of highly relevant regulations.

Table 14.1 Examples of important regulations in the biotechnology and pharmaceutical industries

21 CFR Part	Area of regulation
11	Electronic records and signatures
50	Protection of human subjects
54	Financial disclosure by clinical investigators
56	Institutional review boards
58	Good laboratory practice (non-clinical)
110	Good manufacturing practice (food)
210/211	Good manufacturing practice (drug)
312	Investigational new drug application
314	New drug approval
600/601/610	Biological industry
606	Blood industry
700/701	Cosmetic industry
807	Establishment registration and device listing
812	Investigational device exemptions
814	Premarket approval of medical devices
820	Medical device, quality systems regulations

Some examples of the content of these regulations are:

- **21 CFR Part 58: Good laboratory practice for non-clinical laboratory studies**

This section contains the required regulations and information for performing non-clinical laboratory research, which will support later applications for research or marketing permits under the FDA. It includes organization and personnel, facilities, equipment, testing facilities operation, test and control articles, protocols for and conduct of a non-clinical study, records and reports, and disqualification of testing facilities.

- **21 CFR Parts 210 & 211: CGMP in manufacturing, processing, packing, or holding of drugs and finished pharmaceuticals**

This section contains the required regulations for the minimum good manufacturing practices necessary to produce pharmaceutical drugs, which will be administered to humans and other animals. These include regulations for organization and personnel, buildings and facilities, equipment, control of components and drug product containers and closures, production and process controls, packaging and labeling control, holding and distribution, laboratory controls, records and reports, returned and salvaged drug products.

The Quality Unit

The Quality Unit is an entity at a pharmaceutical company that fulfills both the quality assurance (QA) and quality control (QC) responsibilities. The Quality Unit is independent from production to prevent conflict of interest.

Quality assurance (QA) can be defined as the **regulation of all activities leading to products or services**. It encompasses oversight in areas related to materials, facilities, equipment, design, processes, production, cleaning, records, reports, corrections, prevention, among others. Some of QA's activities and responsibilities include release and rejection of products and materials, creation of production documents and stability programs, and management of deviations, internal audits, customer complaints, recalls and trends (e.g., product and facility).

Quality control (QC) can be defined as the development of test systems that **ensure individual products and services are produced to meet expectations**. The QC unit must ensure appropriate test facilities for chemical, biological and microbiological testing. Examples of QC activities and responsibilities are product receipt and testing (identity, potency, purity, safety and shelf life), test method qualifications, creation of approved methods and implementation of test controls, and investigation of out-of-specification (OOS) test results. If a result is out of the accepted range it is said to be OOS and a thorough investigation must be performed. The outcome of the investigation will lead to either batch acceptance or batch rejection.

Quality management

Several quality management systems have been developed and a few examples are briefly mentioned below:

1. A commonly used industrial standard system is provided by the International Organization for Standardization (ISO), which has members in at least 157 countries. ISO 9000 is a commonly accepted quality management system.

2. Lean, Six Sigma and Lean Sigma organization of manufacturing are other examples of quality management systems. The focus in Lean manufacturing is to eliminate waste by removal of non-value-added activities. Six Sigma employs a variety of tools to decrease defects and is highly customer-oriented. The integration of methodologies from Lean and Six Sigma has resulted in Lean Sigma and this tool has gained widespread acceptance.
3. Process Analytical Technology (PAT) includes "built-in" or "embedded" quality control systems for designing, analyzing and controlling manufacturing quality and specifications. PAT is a very complex set of tools that uses integrated risk analysis dependent on the application (chemical, physical, microbiological, mathematical and others). PAT provides advanced manufacturing intelligence through continuous monitoring and improvement of operations.

Deviations and preventative/corrective actions

When something unexpected happens during a manufacturing process it is called a "deviation." A deviation gives rise to an investigation regarding its nature and cause. The purpose is to identify the source(s) of the deviation, to control it and ideally to eliminate it. Investigations may be simple or more complex depending on the root cause(s). All deviations must be reported and documented as soon as possible in order to quickly begin correction. Deviations are responsible for much additional administrative work. A well-written and thorough investigation pin-points the true root cause of the deviation, so that effective corrective and preventive actions (CAPA) can be implemented. Implementing and monitoring the effectiveness of CAPAs are part of the regulatory expectations.

Audits/inspections

Depending on the type of deviation that has occurred, an audit (visit by regulatory authorities) may be required. This is referred to as a "for cause" inspection/audit. Companies are also inspected to receive approval for commercial manufacturing of a product, referred to as a pre-approval inspection (PAI), or as part of the routine biennial inspection requirement. In an audit situation, company officers will be asked questions about their quality systems. Process and product deviations are often scrutinized and regulators want to know the root causes and any potential product impacts. Auditors serve as an additional reinforcement of compliance. They can be satisfied by proof of improvement or planned improvement. Routine internal audits performed by company personnel are also a regulatory expectation and help a company address inefficiencies to improve its quality systems and meet the current GXPs requirements.

> The outcome of an audit has significant implications for the future of a company

Most regulatory inspections lead to some type of observation. Depending on the findings, these are rated based on risk level for the safety and quality of products. A company needs to address the observations in a satisfactory manner; if ignored it can lead to a warning letter or even a consent decree. As a result, a plant may be shut down or a product may be taken off the market (product recall).

When an audit is occurring in a company it is common to observe a certain level of added stress among the employees. Most employees will try to exhibit their best behavior, avoid noisy activities and be attentive to any request for assistance during the audit. When an audit is over and considered a success for the company, an atmosphere of relief is often observed.

Human error prevention

Even though companies put several control systems in place and attempt to give optimal instructions, there is still a somewhat uncontrollable factor known as human error. Human error is currently a major cause of deviations in pharmaceutical manufacturing operations. Is it possible to avoid human error deviations? Many companies are developing systems to track errors and models of human performance to understand where in the processes the most errors occur, when they occur and which personnel are responsible. Knowing these factors will assist in innovation of new methods to minimize error occurrences.

It has been common to choose a **reactive error control** based on the accumulated knowledge about occurrence of errors and deviations. This means when a person makes a mistake, ways will be found so that this mistake cannot be repeated. It may involve a change in the SOPs, assignment, time management, etc.

A new set of tools is under development which concentrates on **proactive error control**. These tools are based on the anticipation that errors will be made regardless of the thoroughness of an SOP. Instead of an improved protocol, an extra set of control checks may be implemented. For example, for each step in a protocol an extra check may be inserted to evaluate the step that was just performed and the step that is about to be performed. Another example is that of a companion worker who looks "over the shoulder" of a colleague to correct an error before it takes place.

> Pro-active error tools are expected to effectively minimize human error

Anecdote: Impact of out-of-specification test results

Wanda Dowell was hired by a large manufacturing company straight after leaving undergraduate school. Wanda's duties involved quality control of vaccine samples in the early stages of production. According to the standard operating procedures she would oversee and record the data.

One day she noticed an abnormal test result. This was the first time for her. "Maybe I am just tired and things will look better tomorrow. I should not worry so much," she told herself. She signed off on the data. Wanda went home to sleep.

When she came back the next morning, the plant was quiet and production had come to a halt. The supervisor looked at her and said; "It seems like an OOS which occurred yesterday was ignored. The QA team reported deviations downstream and investigations are currently ongoing."

The supervisor knew she was responsible for Wanda's actions and also for her training. It was clear that Wanda had ignored a "red flag" during testing of a product sample. Because she had not reported the OOS result, production continued unaware of any problems and the quality of the large batch of product was put into question. Wanda received training in error prevention and she learned about the implications of a production halt and loss of a batch: time, materials, payroll, loss of sales, etc. Such an easy mistake had such enormous consequences!

With time Wanda grew professionally and personally. She experienced more OOS results and learned that in many cases they represent true process deviations and are not related to laboratory or analyst error. She also learned to adhere to the SOPs and to report any abnormal data immediately.

Fear of failure is common, especially with new employees who would like to impress their supervisors. Supervisors of new employees must learn to anticipate this fact and explain to employees what is most important. New employees must learn to follow SOPs to detail; they must learn to both obey and comply and must never deviate from approved procedures.

Summary of key skills for success

To be successful in contributing to the expected quality you must be able to:

1. Understand CGXP requirements
2. Know which CGXPs apply to your work duties
3. Comply with relevant CGXPs
4. Read SOPs
5. Identify and address any uncertainty
6. Follow SOPs
7. Document your work in *real time* according to regulations
8. Observe and report all data generated
9. Ensure validity of the data
10. Report all aberrant (atypical/unexpected) data and deviations
11. Assess how to avoid deviations
12. Be truthful
13. Take responsibility for your mistakes
14. Be a good example for others

Mindset assessment: Quality in the biotechnology and pharmaceutical industries

Do you work well in a highly regulated setting? Can you comply with regulations? Are you honest with your data? Review the mindsets described below to assess whether you would thrive in an environment with a high level of quality control. If you identify with Mindset 1 you will be most likely to do well in a highly regulated setting. If Mindset 2 is your preference, you would be wise to take another career path.

Question	Mindset 1	Mindset 2
Do you observe and validate all your data?	Yes	Not always
Do you report aberrant data or deviations or do you quietly work to avoid/discard/hide them?	I report these immediately, then work on avoiding them in the future	I like to work quietly to avoid/discard/hide these without interference from anybody

(Continued)

Question	Mindset 1	Mindset 2
Do you thrive in a highly regulated environment?	Yes	No
Are you able and willing to follow SOPs with rigor?	Yes	I like to change things if I feel like it
If you discover uncertainty in a protocol or an SOP what do you do?	I address it and seek to eliminate the uncertainty	I make my best guess and take some chances
Do you document your work?	Yes	Most of it
Can you stay focused when performing highly repetitive tasks?	Yes	No
How do you react if an audit is necessary?	I prepare all the documentation necessary and make it available for the auditors	I resent audits and I get nervous. I suspect that my documentation is not good enough (and it isn't)
Do you like perfection?	Yes, I strive for it	No, I like that there is room for mistakes and independence

Chapter

15

ETHICS

SCENARIO AND IMPACT

Good ethics help establish environments in which people can feel assured that a certain moral standard is upheld. Project ethics relate to everything concerning conduct of a project, such as decision-making, documentation and compliance with regulations. Character ethics relate to all personal conduct. Having discipline is part of good ethics and this topic is described in more detail in Chapter 8.

> Ethical standards provide a moral compass for conduct

Many project regulations exist and make it easier to adhere to a unified ethical code. These vary depending on geographical location and project; there are specific regulations for research using human subjects/specimens, animals, plants and microorganisms. Some examples of this are the testing of animals in pre-clinical trials and testing of human subjects in clinical trials. Samples of regulated areas are given below:

1. Important considerations when performing research on **human subjects/specimens** are the handling of personal data. How data are

collected, shared and communicated with others, stored and used is regulated in order to protect individual privacy.

2. **Animals** must be properly cared for, housed and treated according to regulations.
3. For **plants and microbes** there is a so-called "no spread – no harm" policy. Regulations encompass proper care, storage, containment, use, product manufacture and disposal.

Good ethics allows trust in researchers, their methods and conduct. It helps ensure that there is no spread of private information or research organisms. By adhering to set ethical standards society achieves a general level of safety and a belief that things are being done the "right way."

Because character ethics concerns personal behavior, it has widespread influence, including in the realm of project ethics. The work environment is shaped based on people's mode of decision-making, communication style, attitude and interrelation management. Guidelines and regulations for character ethics are less concise than those for project ethics because so much depends on individual perception. When good character ethics are upheld, people will act in manners deemed appropriate by the majority, thereby creating an environment of "knowing how to behave and what behavior to expect from others."

IMPORTANT KNOWLEDGE

Project ethics

Decision-making and documentation

Professional decision-making needs to be a function of observation, listening, careful analysis, understanding, honesty and knowledge of the effect of the decision. In research, much havoc can be created if any of the abovementioned parameters are ignored. As examples, if a data point is ignored or an observation slightly skewed due to a wishful outcome rather than the actual outcome, incorrect conclusions can be made and can lead the project off track. Data should be observed with curiosity and careful analysis.

> Data always tells the truth

Experiments, data, data analysis, conclusions and decisions must be documented frequently, concisely and truthfully. Example guidelines for good notebook procedures are found in Chapter 12.

Research with human subjects/specimens

Working with human subjects/specimens is exciting, especially when considering the enormous implications it may have for the common health and well-being of mankind.

Human subjects/specimens are defined by:

1. Humans
2. Bodily materials from humans (examples are cells, blood, urine, tissues, organs, hair, nails)
3. Residual diagnostic specimens from humans (for example, tissue obtained through patient care)
4. Private information about humans (examples are medical information, which can be readily identified with individuals, research or cell lines that can be associated with individuals).

There is specific regulatory oversight concerning research using human subjects/specimens. The guidelines and regulations vary depending on the country in which the research is performed. In the United States, Institutional Review Boards (IRB) determine whether or not subjects and specimens can be used for research and whether or not consent from the individuals providing the data or specimen is required. *As long as the data cannot be associated with a living person or living relatives, or if the data are publicly available, the IRB may give approval for use of the data or specimen.*

Research with animals, plants and microorganisms

Animals, plants and microorganisms should be contained in proper and safe fashions, so that harm to the organism and spread of the organism are avoided. As an example, plants may be bred to produce higher yields or sustain drought, and may eventually be allowed into crop production once they have been deemed safe by regulatory agencies and accepted by the community. Another example is the use of microorganisms to produce large amounts of beneficial compounds. These are typically contained in large fermenters and only the product, not the microorganism, will ever leave the site of production. In addition, any organism that is considered a genetically modified organism (GMO) is under specific regulatory oversight.

Character ethics

Communication style and attitude

Good character ethics requires striving to behave in the best possible way. The communication style and attitude of an individual is of utmost importance. This includes both oral and written communication. People who are able to be pro-active, positive, accurate, appropriate, helpful, kind and who display good listening skills (see Chapter 6) contribute to a positive atmosphere.

> Attempt to use a positive communication style

E-mail is one medium through which much misunderstanding occurs. Depending on the mindset of the receiver, the language in an e-mail can often be perceived differently than it was meant. It is best not to let frustration caused by e-mail turn into negativity. Instead, it is recommended to call the parties involved or plan a personal meeting to discuss the different perspectives as soon as negativity is perceived in an e-mail message.

> Negativity will get you nowhere but a step back

Another important matter is which types of messages to relay by e-mail (and other printed material). Certain issues should not be disclosed in e-mail messages. These relate to any legal, confidential and personal matters, especially matters of ambiguity. Personal meetings are the optimal forum to discuss such topics.

> Avoid e-mail correspondence about legal, confidential and personal matters

Crisis management

Dealing with crisis in a pro-active and appropriate manner is a great art. Managers who remain calm, patient and emotionally detached from the matter of discussion are likely to remain objective and realistic. Good communication and listening skills are essential to solving a crisis, as are knowledge about company rules and general laws.

Harassment and discrimination are common crises that can occur in a multitude of areas: Sexual, gender, ethnic, religious and many others. All are inappropriate and the best course of action is to attempt to stop the occurrence immediately.

> Know company policy and report harassment

A key factor to full resolution is concise documentation of the event(s). Important elements include:

1. Date and place of event
2. Classification of event
3. Parties involved in event
4. Description of event
5. Impact of event
6. Signature.

Crises may be reported to a supervisor and the human resources department in the company. As soon as the event is reported, it is the responsibility of the person who accepted the report to take further action. This action may be determined partially by company standards but also by common laws. If the company has had a high incidence of harassment/discrimination events it may be time to re-think career choices and maybe plan to change workplace.

Dress code and hygiene

Employees should be compliant with a professional dress code (see Chapter 3, Professional conduct during the job application process). They must be proper and clean.

Interaction with outside constituents

While traveling to meetings and conferences the time spent is considered work time. The dress code should still be professional, and project and personal conduct should be in compliance with the company policy. Traveling for the company is neither a vacation nor a party, rather certain deliverables are expected. Excessive drinking, spending or partying should be avoided.

> Consider yourself as an ambassador for your company; your behavior is monitored closely and an example to others

Upon returning to the company, reimbursement requests should be honest and reasonable. Usually there are no reimbursements for alcoholic beverages or leisure events.

Anecdote: Crisis

Lina had just been hired. She worked in the lab and was part of a large group of scientists. At the annual Holiday party, she was introduced to the executive staff. She was very proud to be part of the company.

There was one thing though, that bothered her. The VP of research kept trying to flatter and impress her. He put his hands around her shoulders and then on her knees. She kindly pushed him away, but he tried again. Finally, she left the party. The next day he was in her office. "Let's go for dinner tonight, we need to discuss your research," he said. Lina knew it could have negative consequences if she said no. She agreed to talk about research over dinner but was aware that there could be a hidden agenda. There was.

Later that week, Lina confided in one of her older co-workers, Ann. Ann knew exactly what Lina was talking about. Together Lina and Ann went to the HR office to talk to the HR director. It turned out that Lina's case was not at all exceptional. This was not the first complaint, but because of Ann's care for Lina and Lina's courage to discuss the matter it was the last. The VP was asked to leave.

Summary of key skills for success

To practice good ethics you need to be able to:

1. Make decisions with deliberation
2. Carefully perform research and analyze results
3. Be honest

4. Comply with regulations
5. Respect people's privacy and personal boundaries
6. Keep confidential matters confidential
7. Conduct yourself in a professional manner
8. Know when and where to address legal matters
9. Keep professional and private lives separate
10. Do not tolerate discrimination or harassment
11. Keep thorough records
12. Behave professionally when traveling and representing your company

Mindset assessment: Ethics

Are you a person in high ethical standing? To address this issue you can use the following assessment plan, which describes two different mindsets. Mindset 1 is a highly ethical mindset, whereas Mindset 2 is not.

Statement	**Mindset 1**	**Mindset 2**
I am aware of the definition of human specimens	Yes	No
I follow guidelines and regulations carefully	Yes	Sometimes, if it is not too inconvenient or slows me down
I care about the impact of my actions	Yes	Not so much, you cannot please everybody
When my data is ambiguous I:	Re-run the experiment or tailor a new experiment to eliminate the ambiguity	Take the chance that what looks most probable is true
When I am in a crisis I:	Record it, then seek help	Get angry or try to ignore it

(Continued)

Statement	**Mindset 1**	**Mindset 2**
When I am in the middle of a dispute I will always defend my case in all possible ways, by phone, by e-mail and in person	No	Yes
If I am unsure if I am infringing on somebody's intellectual property, I will send our legal department an e-mail	No, I will not use e-mail, instead I will talk to the legal counsel in person	Yes
When I am at a company party	I conduct myself professionally while having a good time	I party as much as I can because it is a rare occasion
When I travel for the company	I conduct myself professionally while having a good time	I take the opportunity to eat and drink at expensive restaurants and bars, rent sports cars and take a vacation
When I return from travel	I ask for reimbursement related to my professional activities	I turn in all receipts

Chapter

16

GLOBALIZATION

Interacting with collaborators and customers world-wide has become a natural part of daily life in many companies. During these interactions, different cultures provide many challenges related to the environment, company organization, communication style and understanding, work ethics and interpersonal relations.

Differences often become hurdles for effectiveness. They may seem threatening or simply tiring. Managing differences is an important skill and it must be nurtured by simple decision-making.

> Decide to respect differences;
> meet these with excitement, not fear
> and display a desire to understand them

SCENARIO AND IMPACT

Globalization takes place in multiple ways:

1. When **adding new employees** from another culture/country
2. When **collaborating with companies** in another culture/country

3. When **merging companies** from another culture/country
4. When taking a **new job** in another culture/country
5. When **starting a company** in another culture/country.

When people from different cultures and geographical backgrounds work together, there is an increased need to ensure that all employees understand goals, deliverables and company culture.

> Very important things can be learned from people who are **different**

The common differences to be expected can be divided into three main categories:

1. *Incentives to work:* Some people work to live, while others live to work, or work to excel and advance. These differences may affect work ethics.
2. *Boundaries:* Some people are used to high levels of intimacy whereas others are used to operating with many regulations and unspoken laws. These differences may affect feelings of trust and integrity.
3. *Background:* People may come from different educational, economic and political backgrounds. They may be religious or not and their social norms may differ. The differences may cause large variations in values and decision-making.

> All can learn many valuable lessons from other cultures – after all, none of us got everything right from the beginning!

Globalization forces people to cooperate with other people and cultures that are very different from their own. People must make extra efforts to be able to communicate and understand each other in an effective fashion. Improved cross-cultural understanding and communication are direct benefits of globalization. Another asset is the enrichment in both technology and intellect created by the differences. When the differences are fully acknowledged and understood, value can be derived from the integration of the differences. Best practices can be established by comparing the differences in practices and deciding the optimal set. Once best practices have been established, people from all cultures must learn to adhere to them. Globalization, therefore, also offers a means to improve collaboration skills and best practices.

Globalization creates an opportunity for improved communication, collaboration and understanding

Important knowledge

Cultural differences

Cooperating with others, especially when they are "different," is an essential component for professional success. Several things can make this difficult and five main areas are of particular interest.

Organization and leadership

Some cultures are accustomed to hierarchical organizational structures with clear ranks and lines of demand. People in such structures follow orders from "above." Other cultures may employ a matrix organization in which people typically feel more like equals. Similar differences may be found in leadership choices. Some leaders employ much control and rigor and make most decisions independently. Others tend to delegate responsibility, management and decision-making. In many cultures traditional leadership roles have been held predominantly by males and there are large differences between cultures with regards to changing this tendency.

Discipline and work ethics

The level of technical, managerial and personal disciplines vary between different cultures. While some cultures reinforce precision, punctuality and quality, others may put more emphasis on personal behavior and reputation. In some cultures there is a "work hard and play hard" attitude, which means that the work ethics are strong when at work and that it is necessary to have a life away from work too. Other cultures have a tradition of "live to work and one's identity is tied to the profession," which promotes long work hours and pressure to increase professional status. Yet in other cultures "work to live and maintain a job" is basic survival and employees do only what it takes to keep the job.

Flexibility

Strict rules, orders and a high level of discipline increase the level of control. Some employees feel comfortable under these circumstances because they know exactly what is expected of them. Others prefer more flexibility to do things the way they deem correct. This may confer feelings of empowerment and trust from "above." Whether one or the other environment is prominent may have to do with the tradition of work ethics.

Communication style and interpersonal relations

While some cultures employ a very direct and assertive way of communication, other cultures seek to avoid conversational styles that can lead to confrontation of any kind. As a result the word "no" is rarely uttered and it can be challenging to understand when interest has receded and an idea should no longer be pursued. Conversely, some cultures may display a directness, abrasiveness and confrontational style, which may come across as impolite in other cultures. In addition to these differences, there are cultures that have a tradition of collaboration or individualism. Highly collaborative employees may find it challenging to get along with people who are highly individualistic and vice versa. Finally, in some cultures it is common (and important) to greet each other with hugs and kisses. In other cultures this would be a violation of the code of ethical conduct and be seen as highly unprofessional and on the verge of harassment.

Cross-cultural understanding and level of globalization

Whether employees in a company are used to interacting with other cultures or only their own plays a large role in determining how well they are going to adapt to the demands of globalization. In cultural "melting-pots" a higher level of tolerance towards differences may be found. This is due to the need to cooperate, as well as have positive experiences and familiarity with other cultures. Cultures that display the best and quickest way of adapting and harvesting the advantages of globalization have the prospect of being the most competitive.

Comprehension of differences

Understanding which differences are common and how the differences are perceived can assist in promoting tolerance and familiarity of others, their values and behaviors.

The table below lists some of the aforementioned cultural characteristics and how these often are perceived by the home culture vs. how they may be perceived by other cultures:

Table 16.1 Actions and their possible perception

Characteristics	Possible perception by own culture	Possible perception by other culture
Matrix organization	Democratic organization, everybody matters	Socialistic organization, lack of leadership
Female leader	Sign of equality	Weak organization
Punctuality	Discipline	Stressful environment
Lack of punctuality	Laid back attitude	Lack of discipline
Working long hours	Hard-working employees	Employees will burn out, do not manage time properly, do not appreciate family
Working 40 hours per week (full-time)	Working as required, able to take care of other needs after work	Lazy
Strict rules and control	Employees know what to expect and what is expected of them	Non-trusting management, lack of personal freedom
Few rules, little control, much flexibility	Trusting management, employees can thrive and have personal agendas	Lax, disorganized
Direct, assertive communication style	Honest atmosphere, no waste of time by guessing	Impolite, abrasive, uncomfortable way of communicating
Non-confrontational communication style	Peaceful, compromising and pleasant style	Hard to know whether an answer is a "yes" or a "no," could be a waste of time
Collaborative work style	Pro-active, creating opportunity, professional	Bragging, pushy, nosey
Individualist work style	Polite, careful, thorough, proper, professional	Not knowledgeable, lacking initiative, lacking marketing

(continued)

Table 16.1 (continued)

Characteristics	Possible perception by own culture	Possible perception by other culture
Greetings with hugs and kisses	Warm, caring, part of family feeling	Inappropriate, sexual harassment, too intimate, unprofessional
Greetings without hugs and kisses	Appropriate, professional, correct boundaries	Cold, uninterested, arrogant

Creating an optimal environment

It is easy to see how conflicts can be created if people do not understand each other and each other's value systems. Even if they do, they may still not agree with each other. What is important to remember is that everybody can be "right" seen from their own perspective and "wrong" seen from somebody else's perspective.

> The important thing is not judging who is right but finding the optimal solution for all

As understanding about different cultures develops, options to try the new and unusual can appear and help to build bridges between cultures as well as provide an opportunity for innovation. At some point, common ground must be established in order to create an effective work environment embracing all of the involved cultures.

Anecdote: International merger

Per Cision's career had led him to fame and his company had been an enormous success. He was a leader of skill and vision, but most of all he was known to institute many rules and to reinforce them. The productivity in his company had soared and the reputation was built on reliability, solid products

and punctuality. After the big expansion of his company in Northern Europe, he had approached a large successful company in the United States for discussions about a merger.

The US company was led by Joe Easy. Joe was a hands-off type CEO with a big heart and a lot of trust in his employees. His company had done extremely well and the turnover was very low. Employees felt comfortable, trusted and they had a fair amount of flexibility in terms of their schedules as long as the deliverables were met.

Per and Joe signed on the dotted line and made the merger a reality. Due to Joe's relaxed style, he agreed to let Per set the tone for the new company. After all, Joe was the CEO who was going to be superfluous; basically he looked forward to retiring to his cottage at the beach. After the merger was official, the companies in both countries underwent reorganization. The European branch needed only slight reorganization. Executives arrived at the US branch in order to implement the work style used in Europe.

The European branch did not feel much of an effect of the merger except on the stock price. After a while, the stocks slid to a dangerous low due to problems in the US branch. There had been much discontent with the new work style implementation. Workers had to clock in and out, they had to be at work at certain core hours and they only had 30 minutes to eat lunch, even during business lunches. They felt controlled and distrusted and the sentiments had spread to all levels. Many of the employees had left the company and their absence and the rumors they had spread about the new management made it difficult to recover.

Per was very concerned and did not understand what was happening. He called Joe at the beach to seek advice. "What happened?" he asked. "I gave my new employees beneficial long vacations, great bonuses and all the great things we were used to in our European branch, but they did not seem to appreciate any of it." Joe had heard a word or two about the troubles and it was easy for him to explain the situation. "My employees are used to having a great deal of flexibility. They work well under such circumstances. Many of them are so grateful they take home extra work. Once used to the flexibility it is hard to change. People are never satisfied with less of anything unless you give them good reasons."

After more discussion both Per and Joe learned more about the differences between their leadership styles. It became clear that it would have been advantageous not to change things so dramatically and so fast. It was also clear that different cultures respond differently. The European workers were used to great vacations and strict work hours while the Americans were satisfied with the flexibility they could translate into trust and control over their own schedule.

Summary of key skills for success

To successfully adapt to globalization you must be able to:

1. Respect others, their cultures and their differences
2. Accept that your own culture may look flawed if seen from another culture's perspective
3. Accept that your own culture and value system may need adjustment in the global arena
4. Decide that differences are an asset
5. Decide that differences provide an opportunity to learn
6. Decide that differences present opportunities to improve communication and interpersonal relation skills
7. Be excited and interested about, not afraid of, differences
8. Understand that you are not right all the time and that it can take extra effort to work across cultures
9. Understand that proving you are "right" is less important than cooperation in working towards optimal goals.

Mindset assessment: Cross-cultural understanding

How globally oriented are you? The statements below may give you an indication. If you tend to agree with the statements, you are quite used to working with other cultures and find it interesting. If you disagree, you may need some time to adjust to the differences encountered in the global marketplace.

Statement	Tend to	Tend to
I am interested in other cultures and their differences and I respect them.	Agree	Disagree
I think that bringing people together from different parts of the world is important, exciting and enriches what we do.	Agree	Disagree
I know it takes extra effort to help people from other cultures understand my perspective and for me to understand theirs.	Agree	Disagree

(Continued)

Statement	Tend to	Tend to
It is necessary that we interact with different cultures and it has become easier.	Agree	Disagree
I do not have to be right all the time.	Agree	Disagree
I understand that I can be right seen from my own point of view and wrong seen from another's point of view.	Agree	Disagree
Different cultures and people vary in many ways: communication style, professional experience, socio-economical aspects, faith and religion and many other aspects.	Agree	Disagree
It is complicated to interact with people, even more so when they are from another culture. That is why more patience, tolerance and careful thinking are required when the interaction takes place. It is important not to make other people feel uncomfortable in any way.	Agree	Disagree
Conflicts are impossible to avoid. I will do my best to adapt kind and understanding ways of dealing with them.	Agree	Disagree
Working with people and cultures I do not appreciate much is unavoidable. I know that this will create tension. It is still my job to make things work as smoothly as possible. I cannot always have a choice to work with my favorite people. Learning to get along with people I do not like is a greater art.	Agree	Disagree
I will teach other people about my own culture and I will learn from them about their culture. Together we will create the optimal environment.	Agree	Disagree

Chapter

17

EXPECTATION MANAGEMENT

When expectations are met people feel satisfied, and when they are not, they may be surprised and unhappy. Therefore, it is essential to know what to expect and to make sure that the expectations are realistic.

Expectations should reflect reality

SCENARIO AND IMPACT

Both employees and employers have expectations of each other. All companies of a certain size have common procedures in place to help both parties assess whether or not these expectations are met. Such procedures normally include personal performance meetings between the employee and his/her supervisor. Performance meetings are commonly held once or twice per year.

Employees meet regularly with their supervisor to review performance and goals

The opportunity to discuss individual performance gives a way for understanding the specific expectations between the employee and the employer.

The assessment is typically expressed as either "exceeding expectations," "meeting expectations," or "not meeting expectations." The assessment is often tied to changes in the status of the employee, such as promotions, demotions and/or a change in responsibilities and compensation.

Exceeding employer expectations can lead to career advancement

The impacts of good expectation management are several. First, employees and employers get a clear view of their expectations towards each other and whether or not they meet these. Second, employers are able to rank their employees internally to determine who is eligible for change in compensation, responsibilities, organizational structure, etc. Third, employees have the opportunity to set new goals for professional development while obtaining commitment from their supervisor to help do so.

The individual performance review is an opportunity to revise goals and expectations

Most employees will try to meet or exceed their goals in order to get raises or bonuses. It is imperative that the employee knows exactly when he or she meets and/or exceeds the goals. Good supervisors will create work situations in which this is possible.

IMPORTANT KNOWLEDGE

Self-assessment

Before the performance evaluation meeting between the employee and his/her supervisor, the employee often writes a self-assessment document. In this document the employee describes all activities since the previous evaluation. The activities are characterized into different categories so it is easy to see how the employee has spent the time and if the goals, as they were defined when the work started, have been accomplished.

Track and categorize progress regularly

Many people are uncomfortable at first when they have to write a self-assessment document because they confuse a self-assessment with an attempt to market themselves. A self-assessment only contains clear facts about what was planned and should have been done, what was actually done and why.

Writing self-assessments can be time-consuming and it is wise to keep track of performance frequently and categorize the activities, for example once per month. There are many different self-assessment "cookie cutter sheets" and an example format is given below together with suggestions as to the areas that might be covered.

Employee self-assessment form

1. Date
2. Name
3. Position
4. Department
5. Supervisor
6. Date of previous review
7. Goals at previous review
8. What has been accomplished or not accomplished and why
9. Personal and professional development

Example: Goals at previous review

The previous goals should be listed here, including all tangible deliverables. Examples of different goals:

1. *Production:* Produce 100 liter of vaccine
   Deliverable: 100 liter of vaccine
2. *Innovation:* Develop new vaccine
   Deliverable: new vaccine
3. *Creation of intellectual property:* Submit invention disclosure form to Legal Counsel and write a vaccine patent application together with Legal Counsel
   Deliverables: invention disclosure, patent application
4. *Management:* Teach and supervise team of four technicians in vaccine production and release team to perform independent production
   Deliverable: team capable of independent vaccine production

5. *Planning:* Write and distribute technical part of business plan concerning production and use of new vaccine to the Vice President of Research
   Deliverable: technical part of business plan for vaccine production
6. *Publication:* Select an academic journal and write draft of publication. Submit final publication once the patent application has been filed
   Deliverables: nomination of appropriate academic journal, paper for publication, submission of paper (depends on filing of patent application).

Example: What has been accomplished or not accomplished and why?

The actual deliverables are described here as well as any deviations from the original plan of work (as established above):

1. *Production:* 90 liters of vaccine was produced instead of 100 due to contamination problems in the month of August
2. *Innovation:* A new vaccine against xxx disease was found
3. *Creation of intellectual property:* An invention disclosure form was submitted to the Legal Counsel. No patent application has been produced yet due to a sudden increased workload for the Legal Counsel. The patent application will be initiated as soon as the Legal Counsel makes it a priority
4. *Management:* A team of four technicians is now capable of independently producing vaccine
5. *Planning:* The technical part of production and distribution of the new vaccine has been analyzed and described in a business plan. This plan has been given to the Vice President of Research
6. *Publication:* No journal has yet been selected and no publication has been written due to time constraints during the month of August (contamination problems, delay in patent application). Instead, the resolution of contamination was made high priority.

Example: Personal and professional development

This section contains a description of what the employee would like to see happen in the future in terms of new responsibilities and development. Maybe there is an interest to lead a group or a new project? Maybe the employee would like to go back to school for a few classes to learn more about a specific area? Maybe a workshop is being offered, which could be helpful? Maybe the employee should present some data at a scientific meeting? There are many possibilities for both personal and professional development and the self-assessment form is a way of documenting personal wishes.

Performance review

After the self-assessment has been submitted, the employee meets with his/her supervisor to discuss the assessment. Many people are nervous about this meeting, especially if they have not met expectations or if they do not know whether they have met them. It is normal to be nervous and usually it helps if the first meeting is a pleasant one. How the supervisor decides to manage this meeting depends on his/her leadership style (see Chapter 5). The meeting may start with a request from the supervisor about an oral review of the employee's work, or the supervisor may review it while the employee is listening. This meeting is a great opportunity to ensure that the assessments are accurate and entirely understood.

> Make sure there are no misunderstandings about the work you have done

Once the assessment has been reviewed, the priorities, new goals and opportunities in relation to the job can be discussed. The common order of the discussion points is given below.

Performance review meeting

1. Self-assessment
2. Supervisor's feedback
3. New goals
4. Performance improvement
5. Personal and professional development
6. Other issues

The chance to discuss personal and professional development directly with the supervisor is valuable because it is a unique opportunity to persuade the supervisor to give an employee a commitment without any interference from others.

Supervisor's assessment

In response to the assessment and the meeting, the supervisor usually prepares a written report. This describes the relevant details and whether or not the employee met or exceeded the goals that were given as well as defining new goals. The report may contain a recommendation concerning altered responsibilities and areas of potential improvement. The report may be shared with the employee and accepted and signed by both the employee and the supervisor. An example of the content is shown below.

Supervisor's employee assessment

1. Date
2. Supervisor name
3. Department
4. Employee name
5. Review of employee goals and deliverables
6. Review of employee development
7. Expectations (☐ not met, ☐ met, ☐ exceeded)
8. New goals
9. Suggested improvements

Date and supervisor signature

Date and employee signature

The supervisor's review is of essence for the employee's future career. Therefore, it is in the employee's interest to behave in a professional and positive manner similar to what is described in Chapter 3 in order to build a sustainable and solid reputation as a valuable employee.

Reputation

Consistent good performance reviews and high ranking help create a favorable impression of an employee. It is important that good performance *and* professional and positive behavior constitute a daily routine in order to make this a continuous impression.

Positive behavior creates a positive image

On the other hand, negative performance and/or behavior may taint an employee for a long time. It can be very difficult to change a less favorable image.

Demand and promotion

A factor that plays a key role in career advancement is **a demand for particular skills and services**. If an employee is successfully working on a project with high visibility and top priority, management executives are very likely to rank this employee highly. There could well be a disposition towards promoting the person. At the same time, if a person fails to perform in such a project, the visibility is also very high.

Work on top priority projects is highly visible

A person who works hard on a low priority project is less likely to be noticed and to obtain an increase in compensation and responsibilities than a person who works equally hard (or maybe less so) on a high priority project. Working on a low priority project generally carries a risk of re-assignment or a lack of demand for the particular skills and services.

Anecdote: Assessing excellence

Kathy was a hard worker. She was thorough, careful and highly productive. At the end of a year it always turned out she had done a lot of work. This was especially evident when compared to some of her peers who seemed to do a lot less, for example Tony.

Tony always seemed to have done little, but for some reason what he did always had a great impact. Of course it was great for the company, but irritating for Kathy. Because Kathy and Tony worked for the same supervisor, this was especially true. They were focused on different projects. Tony's project was tied to the company's major production whereas Kathy's was more experimental and not a company "life-line."

Kathy wrote her self-assessment. It was very long and contained a multitude of tasks and achievements. She had submitted invention disclosure forms, written publications and had performed numerous experiments during the past year. She was a wonderful manager and her team let everybody know how much they liked working with her. She was expecting lots of praise for her hard work and a good raise. Most of all she hoped for a promotion. It was "her turn," she thought. Tony had been promoted last year.

Kathy met with her supervisor and he praised her hard work. They agreed that Kathy had met her goals, even exceeded some of them. What Kathy did not expect was what was to come next. Kathy was told to work less hard on this project because it was of minor importance. Many resources had been used which could be of better use elsewhere, for example in Tony's group. She was told that she could learn to work smarter, not harder, from Tony by working with him in his group.

After the company's performance review period all employees were compared by their supervisors and ranked. To make things even worse, it became evident that Kathy was not ranked among the best employees. Therefore, she was neither getting a promotion, nor a raise.

After a little while and some distance from the old project, Kathy realized that by her brute force approach she had been productive, yet not a great asset to the company. Basically she had not given herself time to think about the rationale behind what she was doing. She developed trust in Tony and together they took time to analyze the approach they would take before starting their work.

Kathy started liking her work better and it was obvious to all. She was promoted the following year after her performance review.

It is common for people like Kathy to have unrealistic expectations about their career situations. Kathy believes she is doing a great job (and she is) and that she should be promoted. She tends to compare herself to colleagues at similar levels and think that compensation must be "fair," that is, have some equitability. What she is unaware of is that the company may have priorities that are different from hers. Once she understands this fact, her compensation and status make sense to her.

Summary of key skills for success

To manage the expectations between you and your supervisor you need to:

1. Know exactly what is expected of you
2. Know what it means to meet the specified expectations

3. Know when you have not met, have met or have exceeded the expectations
4. Document and categorize your achievements
5. Be able to write a brief and concise self-assessment
6. Be able to communicate well with your supervisor
7. Know whether your supervisor understands what you have achieved
8. Be able to receive feedback in a positive manner
9. Keep a professional attitude even when the topic is **you**
10. Be willing to improve your performance
11. Be willing to develop more skills

To manage your own expectations regarding your career you need to:

1. Accept your supervisor's opinion about you and your project
2. Accept your supervisor's recommendations
3. Know if your project is of high priority to the company
4. Determine whether you want a change in visibility
5. Be realistic about your own expectations
6. Adjust your expectations as necessary

Mindset assessment: Expectation management

Are your current expectations about your career advancement realistic and in sync with your supervisor's evaluation of you and the bottom line in the company? Do you manage your own and others' expectations well? To understand more about your personal situation, the following assessment may be useful. If you tend to have Mindset 1, you can improve your expectation management skills dramatically. If your mindset is more like Mindset 2, you are most likely managing expectations well.

Mindset 1	Mindset 2
I am doing an excellent job	I most likely have areas where I can improve

(Continued)

Mindset 1	Mindset 2
My project is extremely important	As long as my project is of high priority to the company it is important to me
If I do well I should be rewarded	If the company does well and I am part of the success, then I should be rewarded
I have exceeded my expectations	I hope my supervisor agrees that I have exceeded my expectations
I am not quite sure about what is expected of me	I know what is expected of me
I write my self-assessment when I am told to do so	I regularly document my progress
My self-assessment is a chance to promote myself and I make sure it describes all aspects of my personality	My self-assessment is a summary of my achievements
I know myself well and may need to educate my supervisor	I listen to my supervisor to learn more about myself
I do not think I can improve anywhere, I am doing everything I can	I am excited to hear in which areas I can improve
I think I should be promoted if my peers are promoted	It is up to my supervisor to decide when I should be promoted
I deserve a bonus and special recognition this year	I will be grateful for a bonus or special recognition if the company deems it appropriate this year
I do not like to compromise concerning my expectations	I hope my expectations are realistic and if not, I will change them

Chapter

18

CAREER MANAGEMENT

Managing one's career is something only a person can do for him- or herself. It is necessary; it applies to any job, at any time and in any place.

> Managing your career is **your** job

Knowing that a career has to be managed and that doing so is a kind of an "extra job" can help an employee assess his or her current situation. Learning to continuously identify one's own goals and if and how they can be reached in the current job situation is part of good career management.

> Do your goals and the company goals coincide?

This chapter is dedicated to the comparison of employer and employee goals and can be viewed as a continuation of Chapter 2. Several topics from the previous chapters are revisited in an attempt to address both employee and employer perspectives and to elucidate whether a clash or an integration of goals can be expected.

Scenario and Impact

Employers appreciate that employees evaluate their goals and their job situation. For employers, a satisfied employee means a more productive employee. If an employee is dissatisfied and the dissatisfaction is tied to company culture or products, it may be of high importance to the company leadership to correct the issue causing dissatisfaction. On the other hand, dissatisfaction can be perceived as the employee being disgruntled. A disgruntled employee is typically perceived as a problem employee. Avoiding "disgruntled-ness" is essential for a company. Whether this takes employee development, reorganizations or lay-offs may matter less to the employer but may be very important to the individual employee.

> Employers will seek to dismiss disgruntled employees

An employee becomes a key player for his or her own success and satisfaction by identifying personal and professional goals. It is wise to review the situation at regular intervals, for example every six or twelve months, so one knows whether personal goals can be reached and if one is on the right path.

> Know personal and professional desires
> and assess if the current path is the right one

Part of the review includes an assessment of the future prospects. What is the purpose of the project and the work? Is the project going to continue? How is the company doing? What does job security look like at the moment? The answer can be tainted by the level of satisfaction felt by both employer and employee. If prospects are good and the path is right, it may seem like a good idea to continue.

> Review the future prospects

Devising a way to attain such knowledge is a complicated task and it can be done in many different ways. As a general statute one must be able to look at all the components described in this book (business goals, bottom line, leadership, teamwork dynamics, communication culture, marketing policies, level of discipline and ethics, flexibility, acceptance of creativity, level of ambiguity and ambiguity management, intellectual property policies, technology, type of work, quality of work, compliance management, cross-cultural understanding, work style, personnel assessment, prospects for future development, and expectation management) to assess if the mission and culture of the company fits with personal priorities.

> Take time to learn about the work environment before evaluating it

There are some universal steps that can be taken in this process:

1. Establish goals
2. Research resources.

If personal and professional goals have been identified and ways to reach them devised it may be time to meet with a supervisor. Together, the employee and the supervisor can establish whether or not the support and resources necessary are available.

> Take the necessary steps to attain the goals

This "extra management job" may seem superfluous if the only goal is to secure a job. On the other hand it may seem to be extremely important if the goal is to retain a job, or if the employee is a person with great expectations for his or her environment and/or their professional growth and status within the company. No matter what the situation is, it is good to be or become satisfied with the job situation.

Satisfied employees focus on development and opportunity, growth and success. They are likely to stay excited about their jobs and up-to-date in their field. This scenario creates a competitive workforce and adds to the competitiveness of the company, an optimal situation for all parties involved.

> Good career management is a win–win for employer and employee

Important Knowledge

Comparison of company and personal goals and priorities

Finding a job that is a 100% match with personal goals and priorities is a challenge. It is normal for some compromises to be made in any type of job. Knowing what one's goals and priorities are is necessary, but knowing on which of these one can compromise while staying satisfied is even more important.

Important aspects to consider when evaluating a current job:
1. Job requirements, abilities and work style
2. Comparison of employee and company goals
3. Comparison of employee and company image
4. Comparison of employee and company culture
5. Comparison of employee and company expectations

Job requirements, abilities and work style

As a first step, knowing one's abilities and work style (Chapter 2) tells an employee whether or not he or she can become satisfied based on the *type of jobs* the company can offer.

Comparison of employee and company goals

The company goals should be evaluated and compared to personal goals. Understanding **business goals and bottom line** (Chapter 4) will help develop understanding of which kind of **deliverables** the company values. The kinds of **technology and knowledge** needed will become apparent (Chapter 13) and they may or may not be exciting. In addition, the level of **creativity and out-of-the-box thinking** (Chapter 10) that is required vs. the level of repeated activity can be learned. Finally, an impression of how much importance **intellectual property** (Chapter 12) does play and whether one is able to **contribute to this as expected** (Chapter 17) should be assessed.

Comparison of employee and company image

The **company image** may have great importance and determine whether or not the company is one for which a particular person can work and the company's production something of which to be proud. The company **leadership** (Chapter 5) may play a role in this assessment and so may the use of **marketing tools** (Chapter 7). If the company is known for its good **ethics** (Chapter 15) and high level of **quality** (Chapter 14), this may overshadow other less favorable attributes.

Comparison of employee and company culture

If the company culture helps create a pleasant environment it may be because the employees are skilled at good **communication** (Chapter 6) and are good at **teamwork** (Chapter 5). In addition, they may show excellence in a variety of fields such as **discipline** (Chapter 8), **flexibility** (Chapter 9), **ambiguity management** (Chapter 11), good **ethics** (Chapter 15) and **cross-cultural understanding and respect** (Chapter 16). Work style and personal priorities will determine the possible level of satisfaction vs. tolerance in these areas.

Comparison of employee and company expectations

Expectations for oneself and towards the company need to be realistic, as do the expectations of the company with regard to the employee (Chapter 17). Together with company executives, the expectations of **deliverables** (Chapter 5), **personal development** (Chapter 17) and **career development** (Chapter 18) can be established. The ease or difficulty with which this is done will give clues about the level of **flexibility** (Chapter 9).

Strengths and weaknesses

A job description provides information about the specific job requirements (Chapter 3). The actual requirements and expectations are stated regularly by an employee's supervisor (Chapter 17). When this knowledge is combined with an employee's expertise, abilities, work style and work experience, it is possible to identify strengths and weaknesses and to understand whether the current job is a good match, even a temporary match.

Comparison of company requirements and personal abilities and work priorities

1. List employer's job requirements in priority
2. Rank your abilities within job requirements
3. Rank your work priorities within abilities
4. Identify level of alignment between job requirements, abilities and work priorities. List strengths and weaknesses by comparing the order in 1, 2 and 3. If identical or similar, the job is a good match. If large discrepancies exist, the job may either be a match if abilities match and work priorities can be compromised, or not be a good match at all

Analyses of three hypothetical employment examples are given: a technologist, a business analyst and a group leader. Each of these persons has different abilities and work priorities leading to a match, a mismatch that can be turned into a match and a mismatch that cannot be anything but a mismatch, respectively.

Example 1: The technologist (match)

Company's job requirements in priority	Abilities in priority	Work priorities	Comparison of order of priorities	Potential
1. Technology	2	1	1:2:1	Strength
2. Independence	3	2	2:3:2	Strength
3. Project management	1	3	3:1:3	Strength
4. Troubleshooting	4	4	4:4:4	Strength
5. Entrepreneurship	5	5	5:5:5	Strength
6. Follow SOPs	6	6	6:6:6	Strength

Example 2: The business analyst (slight mismatch)

Company's job requirements in priority	Abilities in priority	Work priorities	Comparison of order of priorities	Potential
1. Business understanding	1	1	1:1:1	Strength
2. Analytical skills	2	4	2:2:4	Potential strength, slight reluctance
3. Ability to organize large amount of materials	6 Wants to improve	3	3:6:3	Potential strength if improvement
4. Teamwork	4	5	4:4:5	Strength
5. Interpersonal skills	5	2	5:5:2	Strength
6. Flexibility	3	6	6:3:6	Strength

Example 3: The group leader (mismatch)

Company's job requirements in priority	Abilities in priority	Work priorities	Comparison of order of priorities	Potential
1. Team work	None	None	1:None:None	Weakness
2. Interpersonal skills	None, is not interested	None	2:None:None	Weakness
3. Collaborative behavior	None, has an individualist work style	None	3:None:None	Weakness

(*Continued*)

Company's job requirements in priority	Abilities in priority	Work priorities	Comparison of order of priorities	Potential
4. Project management	1	1	4:1:1	Strength
5. Discipline	3	2 very consistent	5:3:2	Strength
6. Technology	2	3	6:2:3	Strength

Need and options for change

If a person is dissatisfied with elements of a job, then there is a need for change. As a rule, when the change is related to one's own job then there is a good chance that change can happen if the request is deemed realistic and reasonable. If the change involves several other parties, company goals, image, culture or expectations, then the chances are much smaller that the change requested can and will be implemented. If a change is necessary and if the change is not possible to implement, then it is time to switch employers.

> A struggle with yourself may be easier than a struggle with everybody else

Assessing marketability

The time to change jobs may come unexpectedly or may be a conscious choice. It may be because of company reorganization, job dissatisfaction, personal reasons or simply an opportunity too good to ignore. It is a good idea to assess one's own marketability regularly by applying for other jobs periodically. There are, of course, issues of sensitivity and confidentiality, which need to be honored. Once an employer perceives that an employee is looking for another job it is possible that this can be perceived as job dissatisfaction. The wisest approach is to request that the job application is treated confidentially.

> Include a request about confidentiality and be aware that it may or may not be honored

Anecdote: Dynamic career management

Paul was an extremely successful business analyst, a real talent with a positive attitude. This was probably one of the reasons he could get along with almost anybody. The company Paul worked for had started noticing this talent and many of the executives would ask Paul to come to their most important meetings with prospective partners. Paul was promoted quickly and received more and more responsibility. He was also starting to travel more and seemed to become a very busy man.

Suddenly, Paul's mother was diagnosed with a severe and terminal illness. She needed continuous care. Paul's father, who was old and disabled, could not provide for her situation. Paul's job put a strain on his ability to assist his family. At this time in his life Paul had to decide what was most important to him, a high-paced, well-paying job with no time for other matters or a more moderate situation where he could take the time necessary to care for his mother.

Paul talked to the HR director about his situation. Together they developed a plan to allow him to step down to a less demanding job at least for the time period in which his mother would need the most care. During his mother's illness Paul developed renewed affection for his family. He gained understanding for other employees with similar needs to assist their families. Paul became an advocate for others during times of challenge.

Paul's situation is not uncommon. As time goes by new challenges and new demands on time occur. It is necessary to stay flexible and anticipate ways to embrace these challenges. Paul accepted a less demanding job with the same employer, which fortunately was possible. Not everybody will experience this kind of flexibility from their employer. Sometimes it is necessary to find a new job or be without a job for a while.

Personal needs change with time. The most stable situation is one in which both work time and personal time are respected and where new challenges can be met with flexibility. Sometimes this is not possible, but an effort can be made to put the right boundaries in place.

> Manage career **and** time away from the career:
> balance work, play and rest

Summary of key skills for success

To successfully manage your career you need to be able to:

1. Understand that managing your career is something **you** have to do
2. Know what your goals are
3. Know what the company goals are
4. Understand whether or not your goals are realistic
5. Know your preferred work style
6. Be able to identify the core requirements of a job
7. Compare your goals and work style with the job requirements and identify your strengths and weaknesses
8. Assess whether or not you are a good match for the job
9. Take time to evaluate your job carefully
10. Know if you are satisfied with your job
11. Change the situation if you are not satisfied
12. Modify your current job
13. Further your training and education
14. Take another job
15. Assess your marketability regularly
16. Manage your career according to the other demands on your time
17. Make an effort to balance work and time away from work

Mindset assessment: Career management

To assess whether you are practicing good career management, you may consider the following statements. If you agree with these statements, you are most likely managing your career well.

1. I am looking out for myself and my own career. I do not expect anybody else to take responsibility for my career.
2. Before I apply for a job I do some research to determine whether the job requirements are reasonable for somebody with my goals, abilities and work style.
3. I understand that my job can change because the job environment is dynamic. Therefore, my job may not stay the same. It could change to something I like better or worse.

4. I know that job satisfaction has to do with how I view the situation. If I do not like what I do, I can either change my attitude or change my job. Changing my attitude may not be possible unless I take another job.
5. Changing my job situation could involve my own effort to re-educate myself. I am open to new opportunities to learn.
6. I realize that my abilities and my goals are not always aligned. I could be very good at something I don't feel passionate about. Alternatively, I could be unskilled at something I have a strong desire to do. It is up to me to decide what is most important to me and how I can attain my goals.
7. Changing to a new job takes a lot of effort. Before I decide to take such a step I will evaluate my situation carefully.
8. There are always going to be issues that make the workplace less than ideal.
9. I try to stay up-to-date concerning new job opportunities that would be relevant to me. I apply to other jobs regularly to assess my marketability.
10. I try to balance my life at work with my life away from work. Having time for family, hobbies, social events and rest is important to me.

EPILOGUE

I started this book by describing success as a function of satisfaction. I would like to finish by continuing this line of thought, and bring to your attention the **most important secret to success**:

> Gratitude is a primer for satisfaction

Focusing on the privileges in your job (and life) drives gratitude and decreases the likelihood of your becoming submerged in the attributes for which you care less. A choice to enhance the positive aspects and opportunities of your job can influence your surroundings in a beneficial manner, by instilling optimism, enthusiasm and energy, thereby improving your work environment.

As life continues it usually becomes increasingly urgent to visualize one's own potential contributions to corporate goals, life and the world, and to ask the question, "What should I strive for exactly that would make my work worth my effort, for myself and for others?"

For me, the goal is to:

> Attempt to do the most possible good for all while using a minimal amount of resources

I wish you success in achieving your personal goal.

INDEX